# Digital Electronics: A Self-Teaching Guide

*Other Science Books from the Wiley Press*

Ashley, *Background Math for a Computer World*
Ashley, *Human Anatomy*
Crandall, *Pascal Applications for the Sciences*
Dietrich & Wicander, *Minerals, Rocks and Fossils*
Gilbert & Gilbert, *Thinking Metric, 2nd Edition*
Gordon, *How to Succeed in Organic Chemistry*
Houk & Post, *Chemistry: Concepts and Problems*
Kuhn, *Basic Physics*
Kybett, *Electronics*
Moche, *Astronomy, 2nd Edition*
Pearson, *Math Skills for the Sciences*
Ryan, *Basic Electricity*
Sutton, *Ecology: Selected Concepts*

# Digital Electronics:
## A Self-Teaching Guide

**Harry Kybett**
**with Vaughn Martin**

A Wiley Press Book

JOHN WILEY & SONS, INC.
New York • Chichester • Brisbane • Toronto • Singapore

Library of Congress Cataloging in Publication Data

Kybett, Harry.
Digital electronics.

Includes index.
1. Digital electronics. I. Martin, Vaughn, 1946- . II. Title.
TK7868.D5K93 1983 621.381'73 83-21589
ISBN 0-471-88035-3

Printed in the United States of America

84 85 10 9 8 7 6 5 4 3 2 1

# Preface

THE FIRST BOOK, *Electronics With Experiments* by Harry Kybett, introduced the concepts of basic analog electronics. This second book addresses the field of digital electronics, including experiments and self-study quizzes. It is highly recommended that the student study *Basic Electricity* by Charles W. Ryan in this same Self-Teaching Guides Series by John Wiley & Sons before attempting this book. The reader at least should know Ohm's Law and the differences between series and parallel DC circuits. With this prior basic knowledge this book will be very easy to understand.

The practical format of providing hands-on experience makes this book particularly well-suited to the individual looking for a *self-study text* and also to the student in a *technical/vocational school* environment.

No previous working experience with Integrated Circuits (ICs) nor any knowledge of either the binary numbering system or digital concepts is assumed.

ICs fall into two broad classes, digital and linear or analog. This book is only concerned with the digital class. Chapter 1 presents a complete listing and definition of all the terms associated with digital electronics in the order in which they are discussed in subsequent chapters of the book. Chapter 2 is an introduction to ICs and is required learning before the experiments in later chapters can be performed.

Chapter 3 begins with a complete tutorial on interconnecting electronic components and covers the three most popular means of interconnecting them. These three methods are 1) pluggable reuseable protoboards with their associated handy support hardware; 2) wire-wrapping; and 3) the use of printed circuit (PC) boards. Gaining hands-on experience will depend on your being able to reliably construct a circuit and to do so with a minimum of difficulty. Chapter 3 is also a prerequisite for successful completion of the lab experiments.

Chapter 4 demonstrates how problems or statements are translated into actual digital circuits. Chapter 5 shows how basic logic gates are used to solve problems. This is the chapter where the theory and actual practice come together. Chapter 6 deals with flip-flops, the important logic device that is the basis of computer memories and many more sophisticated devices. Chapter 7 consists of shift registers, adders, and arithmetic circuits.

Chapter 8 discusses, among other subjects, analog-to-digital (A/D) and digital-to-analog (D/A) converters. These devices either have analog inputs or outputs and as such are based on the op amp; the D/As and A/Ds are the only nondigital ICs in the book, and are only briefly discussed.

This book is dedicated to my brother, Allen, who passed his CPA exam on the first try. Also, I want to thank Carol Hardin for her patience and cooperation during the writing of this book.

Vaughn Martin

# Contents

## INTRODUCTION

# Why Study Digital Electronics?

THE BASIC DIGITAL PRINCIPLES discussed in this book are not in any manner what might be called truly state-of-the-art, nor are they on the razor's edge or forefront of high technology! These principles (not the technology by which they are realized) have been known for quite some time compared to the relatively new science of electronics. What gives these principles their newly found importance is the technological ability to make them realizable in tangible form.

Most electronic circuits and even memory elements were all built in the past with vacuum tubes, those antiquated, power-consuming, red glowing glass "bottles" that characterize and typify obsolete electronic equipment. But now, with the advent of the integrated circuit (IC), we can realize in a small simple device, running coolly off a single small battery, what would take a room full of tubes to accomplish in the past. Most importantly, this already made-to-order, standardized, convenient building block yields an unprecedented degree of reliability. The MTBF (mean time between failures) for certain SSI (small scale integration) or simpler digital ICs is in excess of 1,000 years. Racks or rooms of vacuum tubes doing the same task had, at best, an MTBF of hours. There were also the factors of expense, power consumption, and air-conditioning requirements. These were all so staggering that the first digital circuit designs in the forms of very simple small computers were mere laboratory curiosities and had little if any practical applications of a commercial nature, until the advent of the IC. Granted, certain differential equations were solved on them, but this specialized use was only for the military's attainment of greater accuracy on data associated with trajectories and related ballistic problems.

To set your minds in the proper perspective and to make you realize how fortunate you are to live in this decade, let us take a look backwards into history. Leonardo da Vinci was an extremely gifted multitalented man with an outcropping of genius which placed him far ahead of his time. His innovativeness was so great that in his writings, which he coded or wrote backwards to avoid being deemed a heretic, he proposed the tank, the helicopter, and the basis of the planetary gear system, just to mention a few of his brilliant ideas. We marvel at this today; but what stopped this energetic man from attaining these dreams? His ideas were pure science at that time; but the means by which they could be realized was technology, and technology was in its infancy.

The digital principles that the vacuum tube designers knew about did them hardly any more good than Leonardo's knowledge did him, since neither could realize his dreams.

Today, everything is different—better! Technology has caught up with and maybe even surpassed pure science. You are therefore the fortunate recipients of the rarely afforded ability to realize your ideas and dreams. You can now take theory, such as the digital principles taught here, and make them immediately and conveniently realizable. That is why this is an exciting field of study for even the most impatient reader or student.

When you look around in the supermarket, you will note bar-graph codes on products and optical readers at the checkout stand to read this digital code. Your watch, the car's instrument panel, television, radio, telephone, credit card, and bank card readers are all based on digital electronics. Everything around you that is electronic in nature has some digital circuitry in it. Therefore, if you want to study the field of electronics, you must also study digital electronics, which has been inherent in all new electronic designs of the last five to ten years.

# CHAPTER 1

# Digital Terminology

**OBJECTIVES:** Upon completion of Chapter 1 you will be able to do the following:

- ☐ Understand the many electronic terms referred to in this book
- ☐ Discuss digital electronics with your friends and/or instructor and use the terminology correctly

THIS CHAPTER, WHICH LISTS and defines digital terms, will aid you in understanding the material in this book. Many of these terms are used interchangeably and are sometimes so closely related that one definition is often merged with another. Many definitions use previously defined terms. We have attempted to build on acquired knowledge and to avoid defining a term using the term itself or by use of a term yet to be defined. The following terms are listed in the order in which they will be encountered in subsequent chapters. An alphabetic list of these terms can be found in the Glossary at the end of this book.

**Digital Electronics.** The broad branch or discipline within the field of electronics dealing with the assignment and manipulation of intelligible data and codes as a function of circuit devices or design techniques in which one of two possible states are assigned. These two states correspond to a device, such as a transistor, either being *on* (conducting) or *off* (not conducting). These circuits operate like a switch and can be designed to make logical decisions. The two assumed states, referred to as logic states, are logic "1" and logic "0".

**Logic 1 or Logic High.** The presence of voltage at a predetermined level at the output or input of a semiconductor logic device.

**Logic 0 or Logic Low.** The absence of voltage or nearly ground level at either the input or output of a semiconductor logic device.

**Numbering Systems.** The system by which we are able to count and relate quantitative data. The numbering system we are most accustomed to is the decimal numbering system which is based on the number ten. Other numbering systems, for example, are the *hexadecimal, octal*, and *binary* system which are based on the numbers 16, 8, and 2 respectively. The largest number expressable by a single digit in any numbering system is one less than that number itself. As an example, in our decimal system we can express "9" as the single largest single digit number. This is $10-1=9$. Likewise, with the binary system, $2-1=1$, so a 1 is the largest single digit expressable number with 0 also a legitimate number.

**Bit.** This term is a contraction of the words *b*inary dig*it*. For example, the binary number $101101_2$ has six binary digits, or bits.

**Gates.** These are the most predominant class of digital semiconductor circuits. Gates are capable of assuming or being assigned one of the two possible binary states of a logic 1 (high) or logic 0 (low). Gates are further characterized by the ability to have certain input conditions produce a predictable output state.

**Truth Table.** This is a systematic and thorough means by which all possible input conditions to a gate are listed along with the corresponding output that will result from each particular input condition.

**AND-GATE.** A gate in which if all inputs are high, a high output will result.

**OR-GATE.** A gate in which if any one of the inputs is high, a high output will result.

**Inverter.** A single input device which sets the output opposite to the input state.

**Exclusive OR.** This special gate can have a high output only if its inputs are not either all high or all low.

*Note*: A NAND, NOR, and Exclusive NOR gate also exist and have truth tables that are just the opposite of the AND, OR, and Exclusive OR gates respectively.

**Integrated Circuit.** A semiconductor device with a combination of interconnected circuit elements inseparably placed on a continuous substrate or single chip of silicon, with small, protruding, regularly spaced metallic pins exactly 0.1 inch apart for connection purposes. These integrated circuits, or ICs as they are more commonly known, are usually encased in plastic or ceramic material. All of the previously mentioned gates are usually realized in the form of an integrated circuit.

**Logic Family.** A collective grouping of integrated circuits with the common factor of design with respect to circuit functions with active devices (transistors).

**TTL (Transistor-Transistor Logic).** This is an example of a logic family with gates and more complex logic functions, realized by the back-to-back placing of bipolar transistors.

**CMOS (Complementary Metallic-Oxide Silicon).** Another logic family in which circuit functions are realized with nonbipolar or MOS (metallic-oxide silicon) transistors. N and P channel MOS transistors are combined for the complementary effect.

**DTL (Diode-Transistor Logic).** An antiquated logic family only rarely used in designs today. DTLs use transistors and diodes, with their inherent characteristic of blocking and permitting current flow, to perform simple logic functions such as gates.

**RTL (Resistor-Transistor Logic).** An even more antiquated logic family in which resistors and their placement in networks, with subsequent voltage dividing effects and cascading of transistor amplification stages, perform logic functions.

**Combinational Logic.** Logic based on the assumption that a certain set of input conditions will cause a predictable output. The mere presence of the inputs in either a high or low state causes the output to be at a prescribed state. The assumption is made that these inputs are not continuously changing.

**Sequential Logic.** Logic in which a certain predefined set or sets of events have first to transpire before the expected end-result will occur. A continuously running squarewave is often used as the driving force so that the experiments based on this logic transpire within a predictable amount of time. The first three chapters of this book deal solely with combinational logic.

**Buffered.** This refers to a special gate with an output that can source (drive) or sink (receive) an extra amount of current, more than the usual unbuffered devices.

**Fanout.** The number of gates within the same logic family that can be driven by a like gate. For example, if a TTL gate can drive ten other TTL gates then it has a fanout of 10.

**Fan In.** The number of inputs that can be connected to a logic input circuit.

**Glitch.** An unwanted pulse that can cause false outputs. The possible causes are numerous, but commonly it is the simultaneous changing of inputs on a multiple input gate in which one input experiences a low-to-high logic level transition while another input at the same time undergoes a high-to-low transition.

**Pull Up Resistor.** A resistor with one side tied to a positive voltage level at or above a potential larger than that which constitutes a logic high; the other end of the resistor is tied to an input on a gate. This insures that the gate's input will always be high until another logic device goes low and in turn pulls this resistor and the gate's input down to a logic 0 voltage level. The pull up resistor convention is helpful in electrically noisy environments in which intermittent glitches could cause false triggering or transition states.

**Floating Input.** An input in which the voltage level is allowed to remain or drift into a range that is neither a logic 1 nor a logic 0 but somewhere in between. This is a forbidden area in which to reside, and often causes hard-to-trace problems.

**Propagation Delay.** The very small amount of time required for the output to change in response to a change in input. This is typically 15 to 20 nanoseconds, or 15 to 20 billionths of a second for TTL logic.

**Karnaugh Map.** A graphic method for simplifying and minimizing logic circuits.

**Boolean Algebra.** This symbolic logic is an unusual type of algebra in which letters and symbols are used to represent statements and their logic connections. This algebra's inventor, George Boole, originally assigned each variable in an equation two values, true and false. Its use has evolved, and today Boolean algebra represents the convention of looking at gates and equating them to open or closed relays or switches, and the logic functions that correspond to this.

**MSB (Most Significant Bit).** In a binary number, this refers to the left-most digit. This digit is twice the value of the one to its right, the next to MSB; and the next MSB is twice the value of the digit to its right, etc.

**LSB (Least Significant Bit).** The bit, or binary digit, to the far right; it is always equal to 1 if a 1 is there, and equal to zero if a binary 0 is there.

**Word.** A collective grouping of bits. The length of a word is up to the circuit designer, although the most popular word lengths are 4 bits (called *nibbles*), 8 bits (called *bytes*), and 16 bits long.

**MSD (Most Significant Digit).** The left-most digit in a grouping of bits or words.

**LSD (Least Significant Digit).** The right-most digit in a grouping of bits or words.

**BCD (Binary Coded Decimal).** A form of numeric representation which expresses each decimal digit as a 4-bit binary nibble. This makes the driving of displays much easier. For example, a number like $3916_{10}$ would be represented as follows:

| | | | | |
|---|---|---|---|---|
| decimal: | 3 | 9 | 1 | 6 |
| binary: | 0011 | 1001 | 0001 | 0110 |

**Register.** A group of memory elements (usually flip-flops) operating in unison to perform a specific task. This task may simply be to hold data temporarily, which is the function of a *buffer register*.

**Shift Register.** The most common type of register. It accepts data and shifts it to the right or left, or adds 1 to the data. Another type of register that merely has a logic 1 circulated around through all its outputs is called a *ring counter*.

**Parallel Loading (Broadside Loading).** When the data is applied to the "D" inputs of the flip-flops. It is also often "gated in" or applied to the input of a multiple-input gate with an enable and inhibit signal on one of the inputs to the gate. This enables controlled loading, or loading under certain specific conditions.

**Serial Loading.** When the data is in serial form and is applied to the "D" input of the first flip-flop in the chain of flip-flops making up the shift register. The data is propagated down the line, bit by bit, on each clock input of the flip-flops. These clock signals can be applied to the first clock input of the shift register. Its output (Q) is then applied to the successive "D" input whose "Q" is applied to the successive "D" input of the next flip-flop, etc. right on down the line. This is called a *ripple counter* because the data "ripples" or propagates down the line serially. Shift registers also exist with only one clock signal applied to all of the clock inputs of the flip-flops simultaneously. This is called *synchronous counting*. Your study of digital electronics will involve microprocessors, and within LSI chips is a device called a *program counter*. This special register combines the features of both a buffer register and a counter. Its purpose is to count and be incremented by one count on each clock cycle, allowing the program to be executed.

**Multivibrator.** A generic term describing a number of different squarewave generators or square pulse generators. Some multivibrators are continuously running, others only put out a single pulse, and others merely change states when they are themselves pulsed.

**Bistable Multivibrators.** As common practice has evolved, they are now called flip-flops. A *flip-flop* is a semiconductor device whose output changes logic states when the proper input conditions are met.

**Monostable Multivibrator.** A semiconductor device that puts out a single pulse in the form of a squarewave.

**Astable Multivibrator.** A semiconductor device that is continuously running and produces squarewaves. This book deals predominantly with the bistable multivibrator type, or flip-flop, as it will be called from this point onward.

**"D" Type Flip-Flop.** A single data input device in which the output corresponds to the logic state of its input when it receives a pulse.

**Triggering.** The process of supplying a pulse to a logic device. Most logic devices we will study from here on will be triggered by a pulse or by a continuous stream of squarewaves. There are two types of devices in logic that are subject to having their outputs change from an input change. These are *edge* and *level* sensitive devices. The first must see a rising or a falling squarewave edge, while level sensitive devices only respond to logic levels. The devices that trigger on the rising edge have no bubble or small circle on the triggering input, while devices that trigger on the falling edge are represented by this bubble or small circle, generally denoting inversion.

**JK Flip-Flop.** This type of flip-flop has two control inputs called J and K. The logic states of J and K control the output.

**Clock.** A term synonymous with squarewave triggering.

**Asynchronous.** An operation that is unlocked or untriggered.

**Synchronous.** The clocking of circuits in which the responses and events in general that occur are based upon certain squarewaves or numbers of squarewaves present.

**Timing Diagram.** A diagram drawn on grid or graph paper to exact scale, in which the relationship of occurring events are compared to the stimulus or squarewave generator driving the circuits.

**Q and $\overline{Q}$.** The two outputs on a flip-flop. These two outputs are always in opposite states if the flip-flop is operating properly.

**Set.** A control on a flip-flop that when active or in the asynchronous mode, controls one of the flip-flop's two outputs, usually the Q output.

**Reset.** Another control on a flip-flop that, like the Set, controls the $\overline{Q}$ output under like conditions.

**Latch or Data Latch.** A special type of flip-flop in which a logic state is retained until a pulse is applied to one input, causing the complementary data to be stored. These devices are used for data storage as opposed to counting and logic functions. This is the most rudimentary form of a memory device.

**Oscillations.** In digital electronics, unwanted switching action caused by either unpredictable switching transistions or induced changing of logic states caused by electrical noise. In any event, this is to be avoided and is done so by the use of a Schmitt trigger.

**Lower Triggering Point.** The voltage level which constitutes a legitimate logic low, above which no triggering action or logic states should change until an upper triggering point is reached.

**Upper Triggering Point.** The voltage level which constitutes a legitimate logic high. Voltage levels below this should have no effect on the switching action. An input signal must reach this level and go low again before the digital or logic device changes states. The time during this transistion state is a function of the amount of hysteresis in the circuit, which is defined below.

**Hysteresis.** As applied to Schmitt trigger switching action, this specifically means that there is a lag time from switching states, either from high-to-low or vice versa, during which either a predefined high or low voltage level must occur before switching action transpires. This prevents intermediate state switching or oscillations in which spurious switching occurs in a region between a truly legitimate logic high or low.

# CHAPTER 2

# Introduction to Digital Integrated Circuits

**OBJECTIVES:** Upon completion of this chapter you will be able to do the following:

- ☐ Identify the pin numbers on a DIP
- ☐ Identify two types of IC packages
- ☐ Explain how to tell if an IC is a CMOS, a TTL, or neither
- ☐ Identify various manufacturers of ICs
- ☐ Define the term *fanout*
- ☐ List the main advantage of Schottky technology
- ☐ Define a buffer or a gate with a buffered output
- ☐ Decide which semiconductor technology, CMOS or TTL, has the wider supply voltage range
- ☐ Identify which semiconductor technology consumes less power at low switching speeds, below 1 MHz
- ☐ List which technologies can drive more gates of its own type
- ☐ Determine which pin is used for which function when using an IC
- ☐ Determine which pins to use when constructing or troubleshooting an IC circuit
- ☐ Identify where the power supply connections to an IC are made
- ☐ List what power supply voltages are needed

# IC Packages and Component Numbering

Most of the commonly available ICs are manufactured in a plastic package known as the *Dual-in-Line* or *DIP* package. ICs intended for military use are packaged in ceramic or nonplastic DIP packages (see figure below). Simple function ICs with less than 13 gates are referred to as small scale integration (SSI); from 13 – 99 gates, medium scale integration (MSI); and more than 99 gates, large scale integration (LSI). An IC chip with more than 1,000 gates, which is quite common today, is referred to as very large scale integration (VLSI).

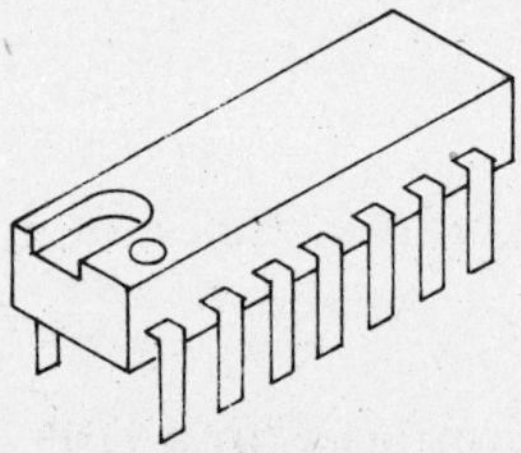

DIP Package

The pins on an IC are numbered. The package is viewed from the top and the pins are counted in a counterclockwise direction.
Start with Pin 1 at the index mark.

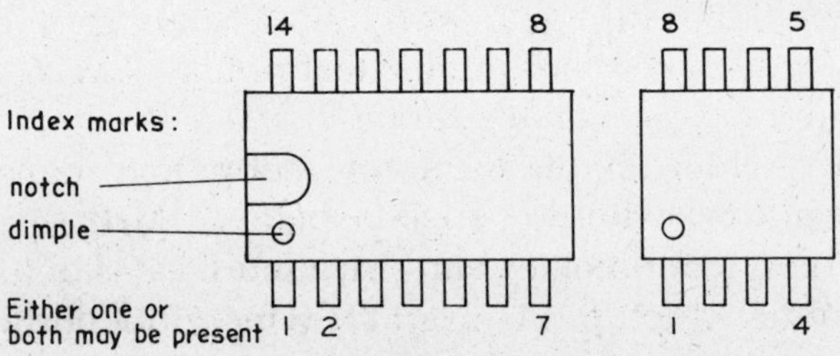


**Q:** Where is Pin 1 on a DIP?

**A:** It is next to the dimple on the IC end by the notch.

Other types of packages are also made. The circular *TO-5 can* and the *flatpack* are the most common. The term "TO" stands for transistor outline and refers to the physical outline of the case. The TO-5 is a metal can and has 8, 10, or 12 pins.

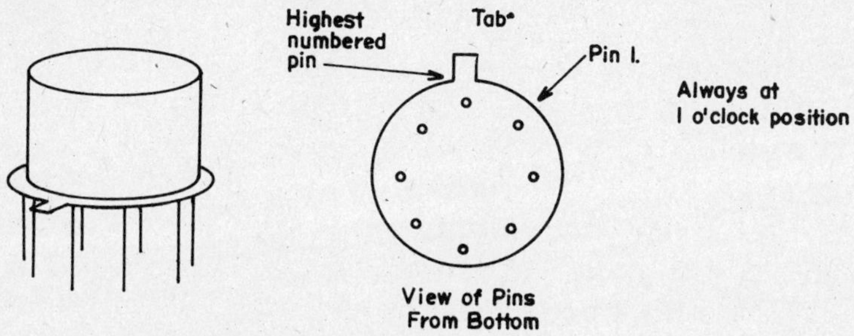


**Q:** What other types of IC packages are found?

**A:** The TO-5 metal can and the flatpack.

The flatpack is not usually found in popular equipment. It is common to military and computer items and is a more expensive package.

All ICs carry an identification number. This can be found on top of the DIP, on the side, or sometimes on the top of the can.

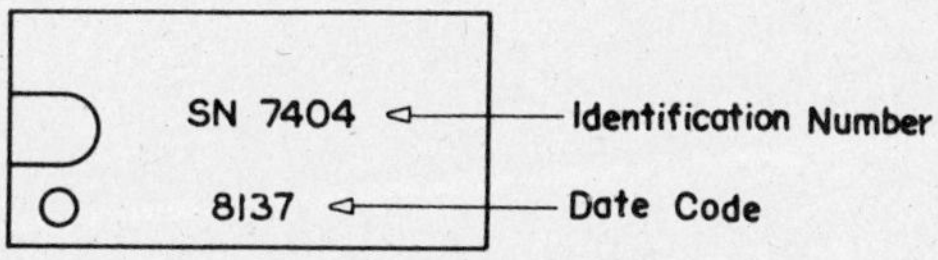


The date code appears as a four-digit number and might be misleading unless you understand it. The first two digits of these four digits are the year and the last two digits are the week within that year. For example, the date code 8137 would mean that the IC was manufactured in the thirty-seventh week of 1981.

The numbering system is slightly different for digital and linear ICs. Digital ICs tend to be grouped into *families*. A family will be identified by its *series number*, and individual ICs within the family are distinguished by the last two or three digits of that number.

**Example 1.** The *TTL* (Transistor-Transistor Logic) digital family has the 7400 series of numbers. The 5400 series is identical in function; the 54 replaces the 74 only to show that it is approved by the military and operates over a wider temperature range. One particular IC in this family contains four AND gates and is known as a QUAD AND gate IC. It is identified by the number 7408.

**Example 2.** Another TTL IC contains two type D FFs (flip-flops). It carries the number 7474.

**Example 3.** The 7400 number itself is used for the QUAD NAND gate IC. Texas Instruments uses an SN prefix and National Semiconductor uses a DM prefix for their TTL ICs.

**Example 4.** The CMOS family uses the 4000 and the 14000 series of numbers. The QUAD NAND in this family is the 4011. The DUAL D FF is 4013. RCA is a major manufacturer of CMOS ICs and they use either CD or CA as a prefix. The "C" stands for CMOS, while the "A" stands for analog or linear, and the "D" stands for a digital IC. Motorola uses the same 4000 series for numbering their devices but they insert a 1 in front of the 4000 series number. Therefore, what might be called a 4013 by one CMOS manufacturer would be deemed a 14013 by Motorola.

**Q:** How can you tell if an IC is a CMOS, a TTL, or neither?

**A:** Look at the series number. If it is TTL, it will have a 7400 or 5400 number. If it is CMOS, it will have a 4000 or a 14000 series number. If it has a different series of numbers, then it is some other IC.

Many identical digital ICs are made by different manufacturers. This is called *second sourcing*. All manufacturers will use the same series number or family number, but each will identify its own product with a trademark or identifying letters in front of the number.

**Example 1.** The 7400 series carries the prefix SN from one manufacturer and DM from another. Thus, the SN7404 and the DM7404 ICs are the same item.

ICs with the same number, but from different manufacturers, are exactly the same device and are completely interchangeable with each other.

**Q:** How can you tell the manufacturers of an IC?

**A:** By the trademark or identifying letters on the package.

Many IC families and individual types have been made. Here we will deal with only a few of the most common and popular, TTL and CMOS. Familiarity with ICs soon removes any confusion about the numbering and type of device one is using.

## Subfamilies of TTL

The TTL (Transistor-Transistor Logic) family has some subfamilies; these are high-speed TTL denoted by an "H" in the part number right after the 74 or 54. For example, a SN74H00 is a high-speed version of the regular 7400 TTL two input NAND gates. An "L" in the same place in the part number denotes a lesser power-consuming device, while an "LS" is a low power-consuming device implemented with Schottky technology, making it both fast and power-miserly. An "S" denotes a Schottky device. Switching speeds, or the time required for the output to change states after the input has changed, is typically 25 MHz (megaHertz), or it will take 1/25 millionths of a second to change. High-speed TTL will switch at a rate slightly greater than twice this fast, and Schottky will switch at a rate somewhere between these two extremes.

**Q:** What is the main advantage of Schottky technology?

**A:** Its switching speed.

The ability of one gate to drive another one is called its *fanout*, the number of inputs that a gate's output can drive. An ordinary TTL gate will drive ten other TTL gates or five sets, resets, or clocks on a flip-flop (a logic device we will study in Chapter 5). An LS or L type TTL device can drive twice the number of similar TTL subfamily gates that an ordinary TTL device can drive. There are also special TTL gates, called *buffers*, that can sink or "draw into them" twice the current a normal TTL gate can: 16 versus 32 mA, to be exact!

**Q:** What does the term *fanout* mean?

**A:** It describes the number of inputs that a gate's output can drive.

**Q:** What is a buffer or a gate with a buffered output?

**A:** It is a gate with the ability of enhanced current drive, typically double the current drive in the TTL family.

# TTL and CMOS Power Requirements

Before explaining where to place the positive voltage, or power, into the IC, let us see what are the acceptable ranges tolerated by these two semiconductor technologies.

TTL, as previously stated, has a 7400 series and a 5400 series of components that are used by the commercial sector and the military, respectively. The military components cost about five times as much as their commercial counterparts. The 7400 series TTL gates operate at 5.0 volts ± 1/2 volt tolerance. The 5400 series parts operate at 5.0 volts ± 1/4 volt tolerance. When performing the experiments, keep in mind that too little voltage will cause improper results, if any; too much voltage will cause irreparable damage.

CMOS (complementary metallic oxide semiconductor) logic operates from a range of either 3 – 15 volts or 3 – 20 volts, depending on the part's suffix. If the suffix is an AE or AF it will only operate over the 3 – 15 volt range. If the part number has a BE or BF suffix it will operate over the wider voltage range. CMOS consumes very little power but is much slower than TTL. It is generally used in 5 MHz or less applications because at about 4 MHz it begins to use as much power as TTL and become inefficient. CMOS will drive 50 other CMOS devices and requires some special precautionary handling. It is usually shipped in a black conductive foam for precautionary measures; electrostatic discharges could cause permanent damage to an IC of this type.

**Q:** Which semiconductor technology, CMOS or TTL, has the wider supply voltage range?

**A:** CMOS.

**Q:** At low switching speeds, below 1 MHz, which semiconductor technology consumes less power?

**A:** CMOS.

**Q:** Which technology can drive more gates of its own type?

**A:** CMOS can drive 50 other CMOS gate inputs.

# Pins and Functions

The internal connections of the actual circuit are brought out to the pins of the package. For a given IC, the pins will be associated with given functions. Manufacturers' data sheets and data books show the pin functions. These reference sources can be obtained from the manufacturers, their representatives, or from most electronic stores.

The data sheets will contain a drawing showing the top view of the pins and small logic or circuit symbols within the package. The functions will be shown connected to the correct pins, as in this typical drawing.

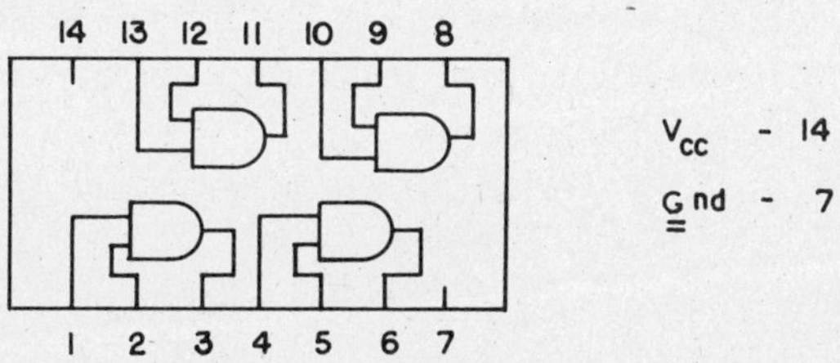


**Q:** How can you tell which pin is used for which function when using an IC?

**A:** Look it up in the manufacturers' data sheets or data books.

Whatever you do—DO NOT GUESS. It is too easy to be wrong, and that is the best way to destroy an IC. When a circuit is designed and published, whether in a magazine or manufacturers' blueprints, it usually contains information about the pins and the functions.

The following circuit diagram leaves no doubt about which pins are used, and this is typically the case with industrial circuit diagrams.

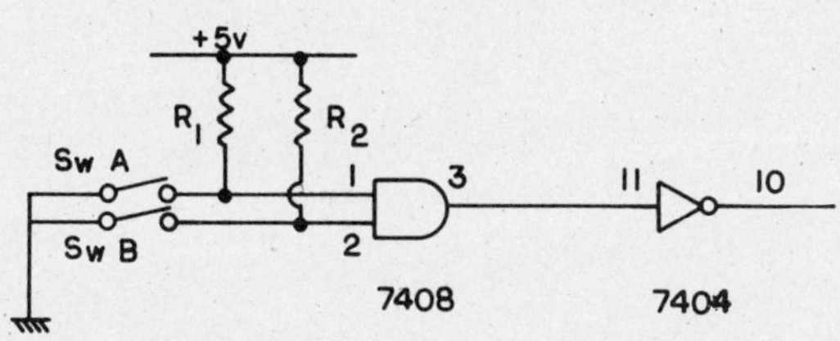

**Q:** When looking at a circuit containing ICs, how do you know which pins to use when constructing or troubleshooting the circuit?

**A:** Consult the drawing; the pins are marked.

**Q:** If the drawing does not have the pins marked, what should you do?

**A:** First, curse the draftsman for inefficiency. Then look up the pins on the data sheets. DO NOT GUESS.

In practice, the pins are often chosen to make the physical layout easier. The circuit is designed first, and after a process of trial and error, the best physical layout is determined. Then the pins are assigned and indicated on the drawing.

In a typical circuit, such as the one at the foot of page 17, it is quite likely that the other pins (those not indicated on the diagram) will be used elsewhere for other purposes.

For example, this may be another part of the same overall circuit.

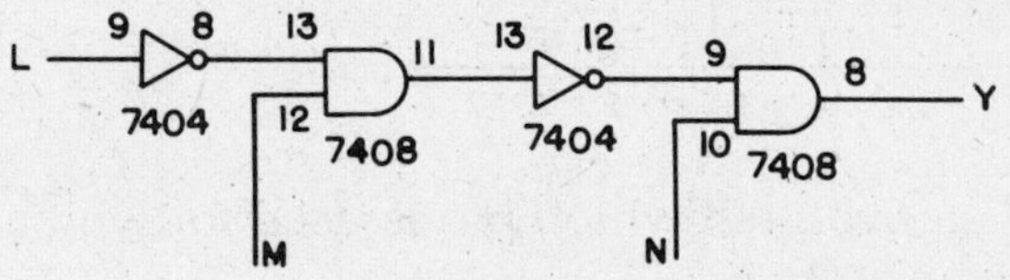


The inclusion of this extra small circuit may cause an original easy layout to be changed. This will be shown in following frames where layout is discussed more fully.

In many IC families, the highest numbered pin is used for the positive power supply voltage, and the pin diagonally opposite is used for ground. For example, here Pin 14 is for +V, and Pin 7 is for ground.

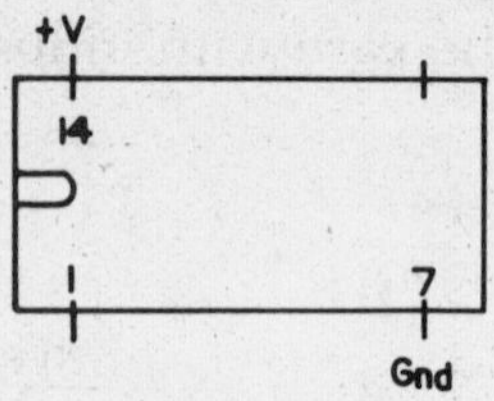

However, this is NOT ALWAYS TRUE. The 7490 is an exception in which Pin 5 is +V and Pin 10 is ground.

In many drawings, such as in the diagrams in the previous frames, the power supply connections and pin numbers are omitted. This is for clarity only. Before constructing a circuit or beginning to check for trouble, determine which pins are for the power supply.

**Q:** Where are the power supply connections to ICs made? (If you don't know the answer, DO NOT GUESS—look it up.)

**A:** Often +V is applied to the highest numbered pin and ground is applied to its diagonally opposite pin. But this is not always true. The only way to be sure is to check in the data sheets.

The voltage supply requirements of ICs depend upon the family type and even the individual type. For example, the TTL family uses +5 volts and ground. CMOS uses 3–20 volts and ground.

**Q:** What power supply voltages do ICs need?

**A:** This varies with family and type. The only way to find out is to check the data sheets. DO NOT GUESS.

When applying power to ICs, several points should be remembered and observed.

1. An overvoltage will destroy most ICs.
2. If the voltage is reversed, it will destroy most ICs.
3. Do not plug ICs into a circuit with the power ON—turn it OFF first.
4. Where possible, use a meter to check the voltage supply.
5. Before inserting an IC into a board, first check all connections carefully, especially the power supply connections.

CHAPTER

# 3

# Digital IC Laboratory Procedures and Exercises

**OBJECTIVES:** Upon completion of this chapter you will be able to do the following:

- ☐ Select the digit that represents a high voltage
- ☐ Select the digit that represents a low voltage
- ☐ Define a *positive logic* system
- ☐ Identify the functions represented by a truth table
- ☐ Draw a simple circuit of the NAND gate
- ☐ Draw the truth table for a NAND gate
- ☐ Draw the truth table for a NOR gate
- ☐ List the output of a NOR gate when it is inhibited
- ☐ List the two useful points for the NAND gate RS flip-flop
- ☐ Calculate time duration
- ☐ Explain the procedure required to replace a TTL IC with a CMOS IC
- ☐ Identify the output of a CMOS 4049 IC with the input floating
- ☐ Determine the output of a CMOS IC when varying inputs are applied
- ☐ Identify which astable multivibrator is distinctly different from the others
- ☐ Write the formula for the frequency of oscillation with an astable inverter in the CMOS family
- ☐ Write the formula for the frequency of oscillation with an Exclusive-OR gate oscillator

# Interconnecting and Wiring Electronic Components

Before any actual "hands-on" experience can be gained in this particular or, for that matter, in any field of electronics, it is an absolute must for the reader to be able to interconnect or put parts together. This chapter solves this potential problem by examining three popular methods of interconnections.

## Overview of Three Interconnecting Methods

The first of these methods is *wire-wrapping*. This method uses wire, its insulation (plastic coating) stripped away at the ends. The end of this wire is then wrapped tightly around a metallic wire-wrap post. The post is usually connected to a pin on an IC socket, or it is part of a whole board of pins spaced 0.1 inches apart on 0.3-inch centered rows.

The second interconnecting method uses protoboards, or pluggable prototyping boards, such as All Circuit Evaluators, produced by AP Products, Inc., a leading manufacturer in this area. These highly durable, white plastic connection areas are filled with numerous small holes placed exactly 0.1 inch apart to accommodate ICs whose pins are also spaced at that distance. The leads of a component or IC are merely pressed down into these holes and the spring action of these holes hold the components tightly. All Circuit Evaluators are used in experiments throughout this book, and it is recommended that you purchase one in order to perform the experiments.

The third and last interconnecting method to be examined is accomplished through the medium of the PC (printed circuit) or printed wiring board. This method has very limited application in experimental and development work due to its lack of flexibility. A PC board has one and only one unique wiring/component interconnecting pattern. This total inflexibility regarding even the slightest changes is responsible for a PC board being as unique as an individual's fingerprints. No two PC board patterns are alike, unless of course they are mass produced.

Epoxy plastics or paper-based phenolic substrates are used for PC boards, with copper uniformly plated or clad to these surfaces on one or both sides. A typical PC board is 1/32-inch thick including the laminated copper coating. A photosensitive resistant agent is usually sprayed on an untreated PC board and the desired interconnecting pattern is then placed over this photosensitized board. When an exposing light rich in ultraviolet rays is shined directly onto the treated photosensitized copper, only the area that has been purposely "blocked out" by the PC pattern remains unaffected. All other parts of the board will be "eaten away" or eradicated during this etching process with a chemical like ferric chloride.

After all the copper has been chemically removed and only the desired PC interconnection pattern remains, the board is drilled. Typically, donut-shaped pads are left, interconnected by straight lines. The centers of these "donuts" are drilled out to accommodate insertion of component leads of resistors, capacitors, diodes, etc. Numerous other special patterns exist such as *DIP* (dual in patterns) and *three lead* transistor patterns.

Each of these interconnecting methods will be examined in detail; Table 3-1, on page 24, summarizes the advantages and disadvantages of each. In summary, none of these systems is inherently better than another. Some are more appropriate for specific applications, and therein lies their popularity. For our use in this book, the All Circuit Evaluator is the best.

### *The Protoboard/All Circuit Evaluator Technique*

The illustrations for this chapter are located after page 31 as a separate insert. It is recommended that you refer to them as you read the text, since they are essential to the following discussion.

Of the three methods of interconnecting or wiring components, the protoboard method affords the easiest way of quickly verifying a principle or circuit concept such as the ones you are about to encounter in this and subsequent chapters. You may want to refer to Table 3-1 again for an overview of this method. The principle is based on the fact that current travels virtually unimpeded in low-resistance paths, such as through metal or wires; but in this case it is through a spring clip that is specifically designed to hold wires with their ends stripped bare of insulation. Actually, all interconnecting media use metal of one sort or another for this purpose. What distinguishes one medium from another is the manner in which this metal is used. The All Circuit Evaluator development board medium is unique because of its use of spring clips. Refer to Figure 3-1 for a cutaway view of a typical spring clip. These devices, along with their white plastic holding boards, collectively make

Table 3.1. A Comparison of Wiring/Interconnecting Methods

| Characteristic Method | Cost | Versatility | Parts Density Per Unit Area | Ease of Assembly | Ease of Changing a Design | Most Appropriate Application | Availability of These Materials to the Normal Consumer |
|---|---|---|---|---|---|---|---|
| Pluggable Prototyping Board such as the All Circuit Evaluator by AP Products, Inc. | Low | Excellent | Medium | Good | Very easy | Quick verification of an idea or circuit | Good, available in virtually all electronic and hobby shops |
| Wire-Wrap Boards | Medium | Fair | Good with the use of a component header | Poor | Moderate | Verification of a system in which numerous circuits are merged into one system | Fair, not as well advertised as pluggable prototyping board materials |
| Printed Circuit Boards | High initial setup costs/ very low mass produced costs | Poor | Excellent if done by an experienced PC draftsman | Excellent | Impossible | For permanent use (not to be changed) in a, e.g., mass produced item whose feasibility has already been proven by the above two methods | Poor, marketing limited to very specific sectors |

up All Circuit Evaluators as shown in Figures 3-2 and 3-3. After the parts are placed on the board during a typical experiment, it probably will look something like the board in Figure 3-4.

The spring clips are bussed together; more specifically, the interconnecting pins that are in a row are electrically common to one another. Note the grooves that both physically and electrically separate these rows of pins, with usually five pins in a row. These grooves are the breaking points where the interconnection of these five bussed pins stop. At the bottom of some of these development boards there are two separate rows of pins. These are common to one another in the lengthwise direction and therefore serve as convenient busses for a certain voltage or ground to be placed there. You may have to put an interconnecting wire between these two busses to ensure electrical continuity (Figure 3-5).

If your budget allows it, there are more elaborate development boards such as the ones shown in Figure 3-6 which can be purchased. The AP Products, Inc. models 102 and 103, on the right of Figure 3-6, are for digital and analog work respectively. They have their own built-in power supplies, as well as monitors in the form of red *LEDs* (light emitting diodes). A wire may be placed in the receptacle below the LED, and if a TTL level potential (3 volt or greater) is present on the wire, the LED will light. On the digital unit, a variable frequency squarewave generator also exists in addition to *bounceless* pushbutton switches. The term "bounceless" means that at switch closure when the contacts begin to physically touch one another, there are not numerous pulses generated as is customary in other switches.

If this very jagged closure of the bounceless switch is placed through a pair of cross-coupled NAND or NOR gates, as we shall study very soon, with pull-up resistors on their inputs, then a perfect switching action will result, with clean logic levels devoid of any glitches or voltage transients present. These special switches can be either 1) built-in; 2) momentary contact pushbutton switches; or 3) two-position switches with logic high and low states that are accessible through the small white plastic receptacle below the switches. Figure 3-7 illustrates a close-up view of a typical receptacle, as well as a typical LED holder with one single spot for a wire to be inserted under the LED itself.

These complete analog and digital development systems naturally have some basic design differences. The digital unit only has a +5 volt power supply as opposed to the analog unit which has a +5, +15, and −15 volt (triple output) set of power supplies. The analog unit also has a meter to monitor voltage. The digital unit has no real need for this meter since in digital electronics, as we will shortly discover, a voltage is either high or low. In a digital circuit an intermediate voltage level should not exist, and if it does exist, it is an immediate source of trouble.

In addition to the ease of interconnections, the next appealing feature of the protoboard is the vast array of products that have been developed to accommodate it. For instance, there are pre-cut and stripped wires in kits assorted by lengths and color coded (Figure 3-8) used on IC clips ranging from small 8-pin up to 64-pin clips (Figure 3-9). Figure 3-10 shows how such a clip would fit over an IC and how a clip-on device replacing the old-fashioned alligator clip can make firm electrical contact with the IC clip. Figure 3-11 shows the variety of these single element clips that are available. In summary, a vast variety of products exist to support these All Circuit Evaluators (Figure 3-12).

### *The Wire-Wrapping Method*

The theory behind wire-wrapping[1] is really quite simple and straightforward. The task consists of removing the insulation from the wire and spinning or wrapping the bare portion around the wire-wrap post. The post is usually a 25 thousandths of an inch diameter square protruding from the board's underside, although it does not necessarily have to be a wire-wrap board. There are IC sockets, connectors (Figure 3-13), headers (DIP sockets with pins coming up to accommodate component leads), and even certain two-lead components like LEDs with wire-wrap post leads upon which a wrap may be made.

In the small engineering development lab, or within your own workshop, a limited degree of sophistication may be present. Usually the wrapping tool looks like a fat pencil (Figure 3-14), does wrapping on one end, unwrapping on the other, and has a slit in the middle in which to place the wire. The wire is then given a quick tug, stripping away the insulation. Further sophistication can be achieved with electric handheld guns (Figure 3-15) that now have bits that automatically strip away the insulation to a predetermined length. The operator then presses the trigger which twirls the bare wire around the wire-wrap post with extreme neatness and with a precise wrapping tension applied.

The wire used by most experimenters and hobbyists is almost always 30 gauge or 30 AWG. The *AWG* stands for American Wire Gauge and #30 has a 100.5 circular mil cross-sectional area. It is this cross-sectional area that determines wire resistance and in turn its current-carrying capability. A general approximation is that for each three numbers you

[1] *Wire-wrap* is a registered trademark of Gardner-Denver Corporation.

go down on the AWG chart you experience a doubling in cross-sectional area. Therefore, one would expect a #27 wire to have double the cross-sectional area of a #30 AWG wire. Actually, it is 201.5/100.5 circular mils for a 2.005 ratio; this is indeed nearly a doubling of the cross-sectional area.

There are numerous types of wire-wrap insulation; Kynar[2] is the most popular with hobbyists because of its ease of stripping. The wire's insulation is no greater than the thickness of the wire itself nor any less than 80% of this dimension. The smaller the wire the more it will stretch.

Wire-wrapping technique is important for a successful (highly reliable) electrical connection. There are essentially two kinds of wraps, the standard wrap and the modified wrap (Figure 3-16). The standard wrap consists of the bare wire being wrapped around the post and the insulated portion being wrapped around the post only a fraction of a revolution, as opposed to the modified wrap which has the insulated portion of the wire making between one and two full revolutions.

The modified wrap is a bit more secure and is recommended for smaller size wire such as the #30 AWG wire the hobbyists use. Figure 3-17 shows the acceptable method of a modified wrap as well as unacceptable methods, and explains why the latter are unacceptable.

Now for the wire wrapost[3] or terminal. This piece of hardware is usually square; however, there are oblong and rectangular shaped posts. These are less popular and effective because of the *pigtail* that typically results. A pigtail is the end of the wire that does not conform to the shape of the post but, rather, has a tendency to stand out. There are three rules of thumb with respect to wraposts. These are: 1) the wrapost should not be less than one wire diameter; 2) the wrapost width should not be more than 2-1/2 times the wrapost's thickness; and 3) the maximum wrapost width should not be more than three times the conductor's diameter.

The wrapost should also be long enough to accommodate two wire-wrap connections. A quick rule of thumb for determining the number of wraps is to divide the AWG number by four; therefore, a #30 AWG wire should have approximately 7-1/2 wraps, with the first and last wraps not counting because they are not involved in the bond determining qualities of the wrap.

The wraposts themselves are commonly made of one of six materials: 1) berryllium copper; 2) phosphor bronze; 3) half hard brass; 4) copper nickel; 5) nickel silver alloys; or 6) plated steel (Figure 3-18).

[2] *Kynar* is a registered trademark of Pennwalt Corporation.

[3] The term *wire wrapost* is used interchangeably with the term *wire-wrap post*.

In summary, you now know all about wire-wrapping. Along with this technical knowledge here are some tips on technique to help you wrap more effectively. Try to wrap one end of the wire on the lower side of a wrapost with the other end on the high end of another post. This is called the *daisychain* method and is illustrated in Figure 3-19. This technique is most appreciated when a wire has to be lifted off an IC pin, because using this method never requires more than one wire to be removed. If, on the other hand, all connections to a point in question were made on the bottom of the wrapost, then a great number of wires would have to be removed and chaos would result.

Here are some additional tips. Examine any wrapost being replaced for excessive damage, taking special care to see if the wrapost is starting to crack and pull away from the board. Insert the wire as far as it will go up into the bit or wrapping tool. Always dress the wire in a wrapping direction. Do not ever try to reapply the wrapping tool to a connection that has been poorly wrapped. Unwrap it and start again! Do not use pliers of any kind on a wrapost; use an unwrapping tool. Do not use the bit and sleeve of the wrapping tool as a pry. Lastly, if possible, do not put over two wrap connections on any one single post.

If you begin working with this interconnecting medium and discover you like it better than working with All Circuit Evaluators, you will need to know about some of the following very handy wire-wrap products. First, the problem I personally have with wire-wrapping is that I constantly have to flip the board over to see where I am; but if you use the identification labels as shown in Figure 3-20 your frustration will be alleviated as well as your tendency to make errors. Another handy item is the IC extractor, of which there are two types: the very fancy kind (Figure 3-21) and the simpler, more moderately priced unit (Figure 3-22). But probably one of the handiest products in this area, and definitely a bargain for the price (approximately $3), is the cut-and-strip tool (Figure 3-23). This tool allows you to cut not only the wire but the amount of insulation you decide to strip back to an exact length.

### The Printed Circuit Board Method

The last, and by far the most specialized and demanding, interconnecting medium is the use of PC (printed circuit) boards. This permanent medium also requires a considerable amount of skill and knowledge before even the simplest projects can be begun or completed. The following information is therefore to expand your knowledge; it is not recommended that you attempt this medium until you have gained more working experience with electronics in industry, unless you work under the close supervision of an individual experienced in this area.

PC work is not suited for prototyping like the two previously discussed methods, which were ideal for quickly verifying an idea or concept (Table 3-1). PC layout occurs only after a circuit has first been verified; however, once a correct PC design has been accomplished, it is the most economical and reliable means of mass-producing any product. Refer to Figure 3-10 and note that the squeeze-on IC clip is used on an IC that is soldered onto a PC board, not an All Circuit Evaluator.

The first step in the generation of a PC design is making a very neat schematic diagram from which to work. A schematic diagram is the most basic of all electronic drawings. And, as you probably already know, it graphically represents the functions and interconnections of electronic components represented by standard and easily recognizable symbols.

After the schematic is in hand, the next step is generation of a layout drawing. This is a representation of the actual physical layout, to scale (Figure 3-24). Note that in this layout drawing there are both solid and dotted lines. Another method is to use the red and blue colored pencils to trace in the interconnecting lines. Most PC boards, unless they are very simple, have layout runs on both sides of the boards. You will want to try to make one side use mostly vertical lines, while the other side uses mostly horizontal lines. This will greatly aid in avoiding "painting yourself in a corner" or encountering a situation where you will cross another line or "run" to get to the place you want to go. Just like wires, these runs can be thought of as conductors and should not touch one another or shorting will occur.

After the pencil layout sketch is generated, the actual layout begins. The medium holding the patterns is usually a sheet of mylar translucent or semitransparent plastic. The patterns are pre-cut and peel off to be applied to the mylar (Figure 3-25). These patterns, whether they be DIP (dual in line) packages for ICs or small donut-shaped discs, are represented two or four times larger than their actual size. Therefore, when the pattern is complete, it will have to be photographically reduced to 50 or 25% its actual size. This resulting negative is what is placed over the photosensitized PC board, and the ensuing process, as described in the introduction to this section, will enable a PC board to result. This PC board is not finished until it is drilled. A small drill (Figure 3-26) is traditionally used for this application along with a carbide bit that is often bifilar wound or that has the regular grooves tapered into the bit, not just in the clockwise manner but in a counterclockwise manner as well.

As stated previously, the PC board is a very permanent medium and not designed for changes or modifications; however, if changes have to occur, the trace cutter (Figure 3-27) will help out immensely. The interconnecting copper conductor runs on a PC board are also called *traces*. Once they are cut, with the use of this tool, they can have small wires make jumper connections. These small wires are called *green wires* because of their color; they are solid, single-stranded wires in the 30 AWG range with green enamel coatings. When these wires are soldered, the enamel almost literally evaporates, yielding bare wire capable of taking solder. If you don't want to lay out your own PC board and if you see an artwork pattern in one of the hobby electronic magazines, you can photographically and quite easily lift it off the page. This is done with a kit like the one manufactured by Datak & CG Electronics (Figure 3-28).

A word about soldering: The importance of good soldering techniques cannot be overemphasized. A single sloppy or faulty connection can have drastic results. If you have never soldered before, please read the following very carefully. If you have soldered before, it is advisable to read the following to ensure that you are properly following the fundamentals of the process. The lost time, frustration, and disappointments you will avoid will make reading this all worthwhile.

What is soldering? Soldering is the bonding together of two or more metal parts with a tin-lead alloy (the solder itself). Soldering is the least expensive and most reliable method of interconnecting electronic components. These electronic components usually have wire leads which attach to one another or to a PC (printed circuit) board. This connection is physically strong and provides a good electrical path through which the current can flow. Solder melts at a relatively low temperature (approximately 370° F) to form this connection. The type of solder recommended is the smallest available (Figure 3-29); and note that the solder on the far right is about the diameter of a mechanical pencil's lead.

Heat is applied to solder by the use of a soldering iron (Figure 3-30). There are soldering irons, pencils, and guns. A soldering pencil is the smallest and is recommended for use here. The point (tip) comes in four basic shapes (Figure 3-31). A conical or semichisel is best for small detailed work like we are doing. You will have to guard continually against oxidation, however, or the build-up of corrosion on the hot tip. The best way to do this is to have a damp sponge nearby and wipe the hot tip over the sponge. This dislodges corrosion and cleans the tip. This is followed by applying solder onto the newly cleansed tip. This process is called *tinning* the tip and ensures against future oxidation.

Once your PC board has been made and drilled and you begin to solder in your components, you may want to protect certain more expensive devices like ICs. One way to do this and to aid in the maintenance/troubleshooting is to either install IC sockets or to use Molex[4] pins. These devices can be soldered directly into a PC board just like a PC board IC socket (Figure 3-32). The method by which these break-off pins are inserted and eventually set on a solderable board is outlined in Figure 3-33.

The Molex pins, wire-wrap wire, and associated products, the All Circuit Evaluators and associated products, as well as PC artwork patterns and related supplies are all available from one source:

Digi-Key Corp.
Highway 32 South
P.O. Box 677
Thief River Falls, Minn. 56701
1-800-346-5144
(Call for a free catalog)

Additionally, every integrated circuit and component to perform all of the experiments in this book are likewise available from this same source.

All experiments in this book show incandescent lamps used as visual indicators. LEDs may also be used. These two-lead devices come in various colors such as green, yellow, orange, and red, with red being the most predominant color, and various shapes (Figure 3-34). A red LED has an almost constant 1.8 volt across it as long as it is conducting or giving off light. The LED conducts or gives off light in varying degrees of luminous intensity, or brightness, depending on the amount of current flowing through it. This range of current varies from approximately 3 mA to 20 mA. There is no need to ever have an LED draw more than 15 mA because at this current level the LED is very bright, quite visible even in direct sunlight.

LEDs are usually mounted on the front panel of equipment (Figure 3-35). Care should be taken in specifying LEDs. The main optical properties are determined primarily by the manner in which the emitted light is given off (Figure 3-36) and the shape of the plastic lens (Figure 3-37).

[4] *Molex* is a registered trademark of Molex Incorporated.

Figure 3-1. A Cutaway View of a Spring Clip, the Basis of an All Circuit Evaluator

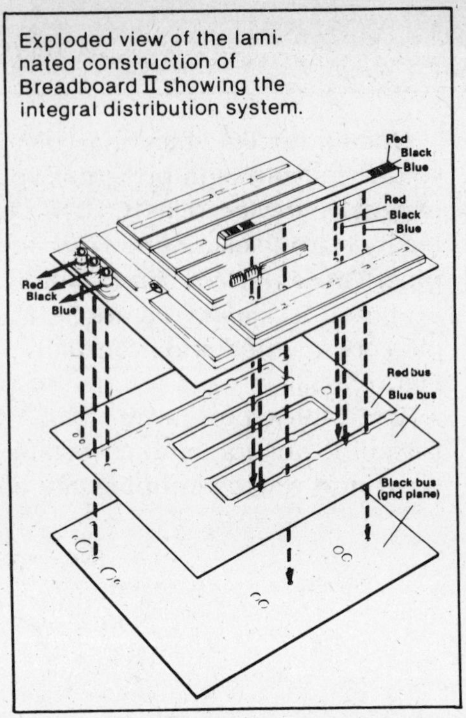


Figure 3-2. An All Circuit Evaluator Exploded View Drawing

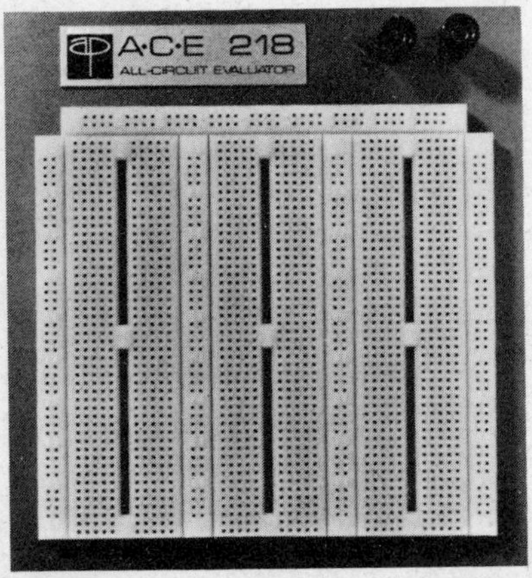


Figure 3-3. An All Circuit Evaluator (Courtesy of AP Products, Inc.)

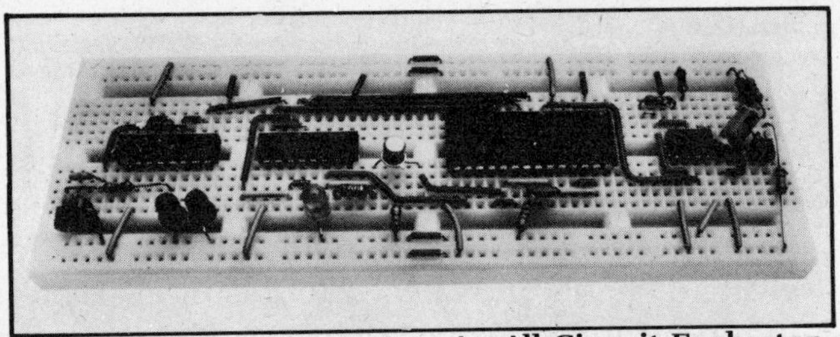

Figure 3-4. An All Circuit Evaluator with a Circuit Built on It

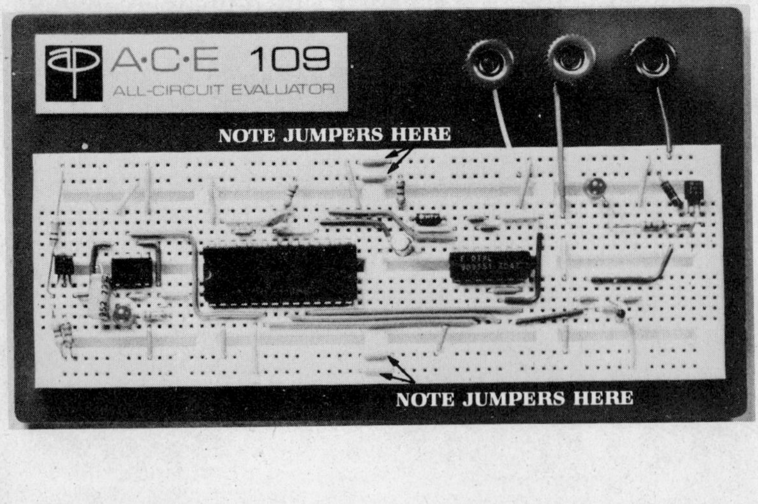


Figure 3-5. Ensuring Electrical Continuity with a Jumper Interconnection

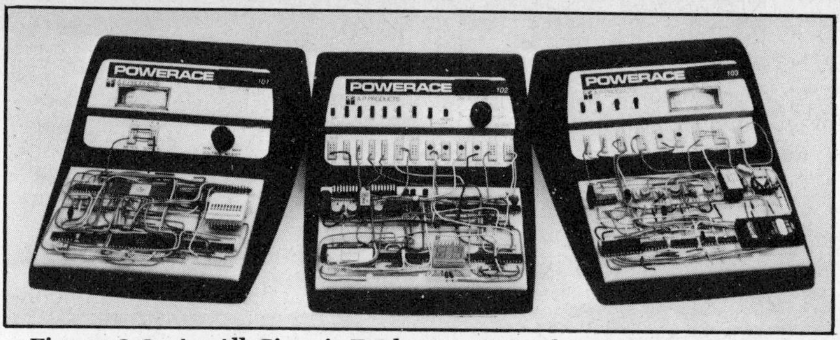


Figure 3-6. An All Circuit Evaluator Complete Development System

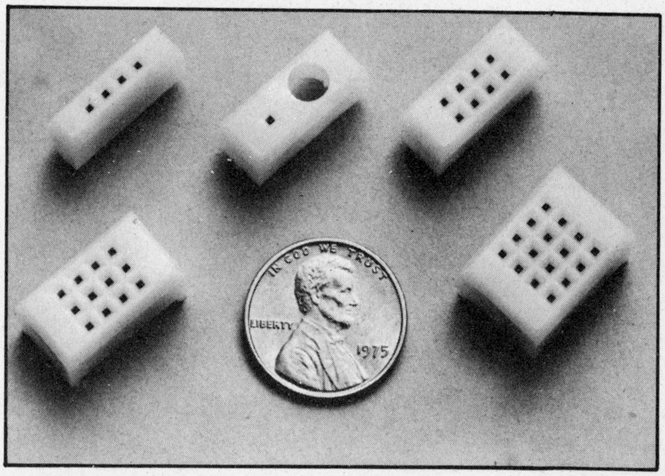


Figure 3-7. A Close-Up View of a Typical Electrical Receptacle

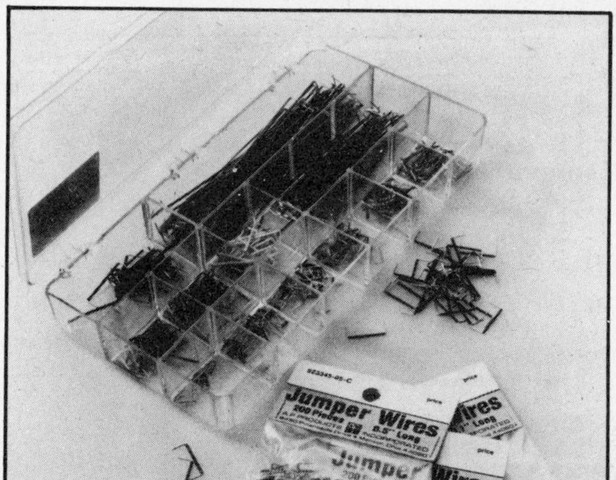


Figure 3-8. An Assortment of Pre-Cut Color Coded Stripped and Formed Wires

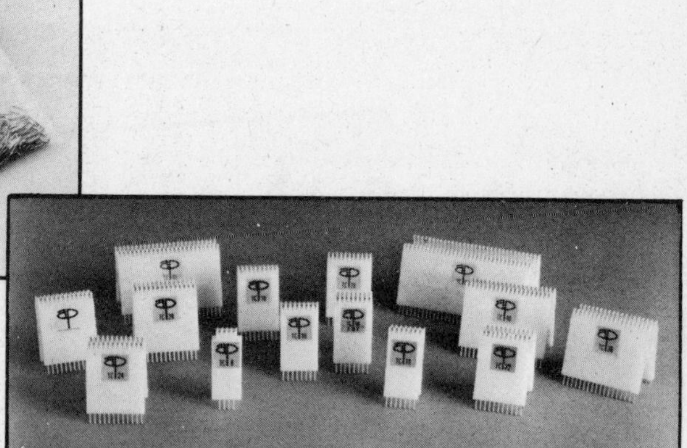

Figure 3-9. An Assortment of Squeeze-on IC Clips

Figure 3-10. A Squeeze-on IC Clip in Use with a Hook-on Type Connector

Figure 3-11. A Variety of Hook-on Connectors

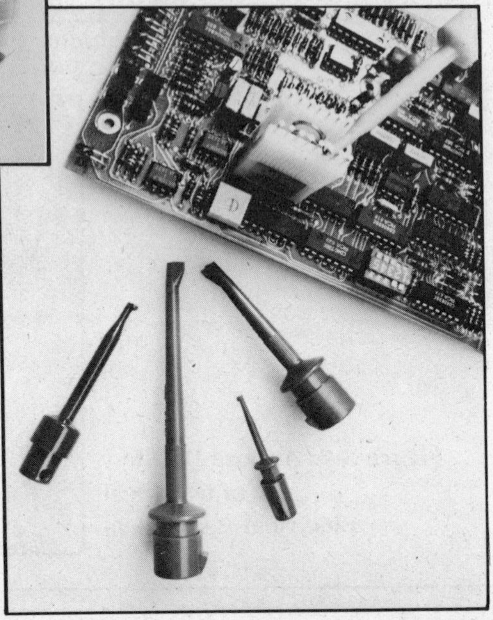

Figure 3-12. A Partial "Family Portrait" of All Circuit Evaluator Products

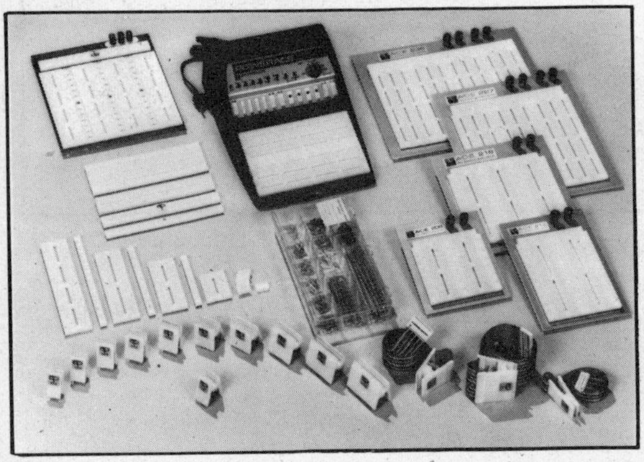

Figure 3-13. A Wire-Wrap Pin Type Connector

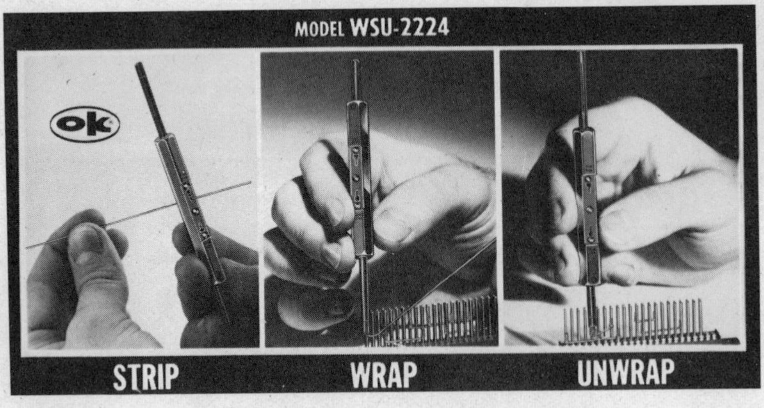


Figure 3-14. A Wire-Wrap Wrap/Unwrap Tool

Figure 3-15. A Wire-Wrap Handheld Electric Wrap Tool

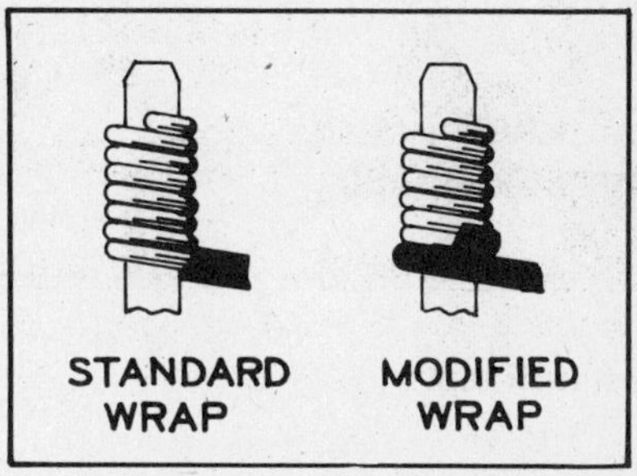


Figure 3-16. The Two Kinds of Wire-Wrap Wraps

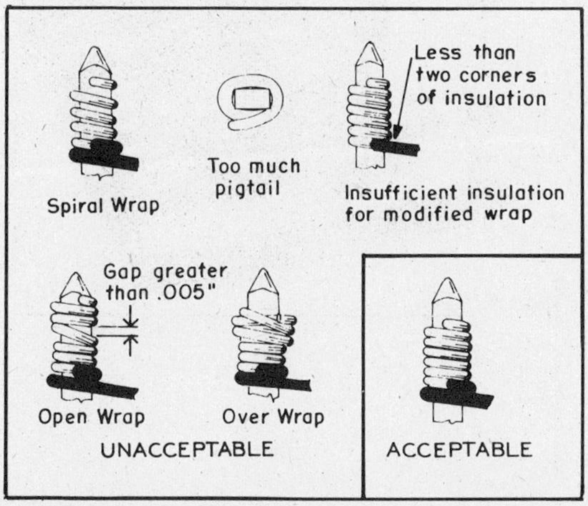


Figure 3-17. Acceptable and Unacceptable Methods of Wire-Wrapping

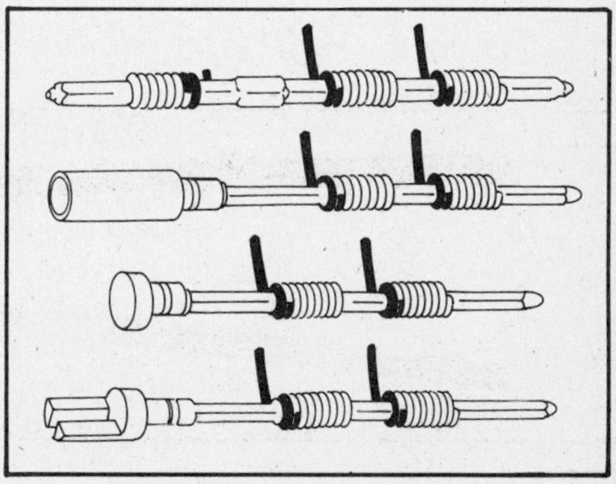

Figure 3-18. An Assortment of Typical Wraposts

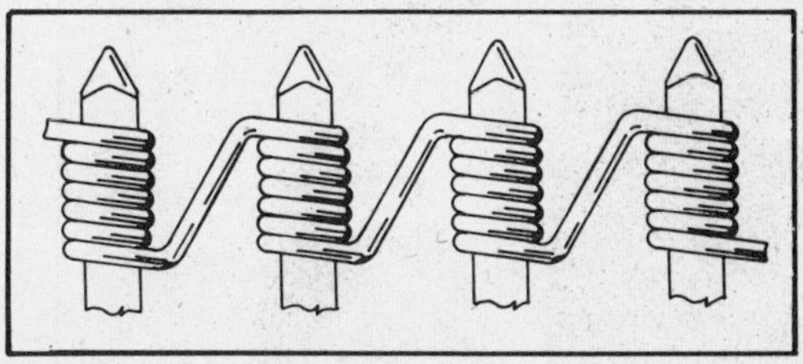

Figure 3-19. The Daisychain Method of Wire-Wrapping

Figure 3-20. Wire-Wrap Identification Labels

Figure 3-21. A Precision Sophisticated IC Extraction Tool

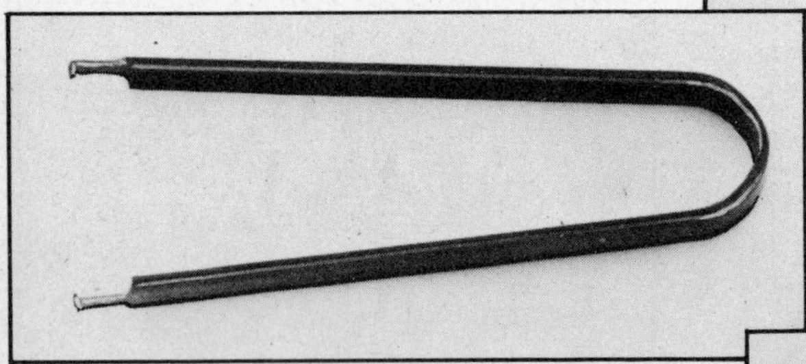

Figure 3-22. A Less Precise, Less Sophisticated IC Extraction Tool

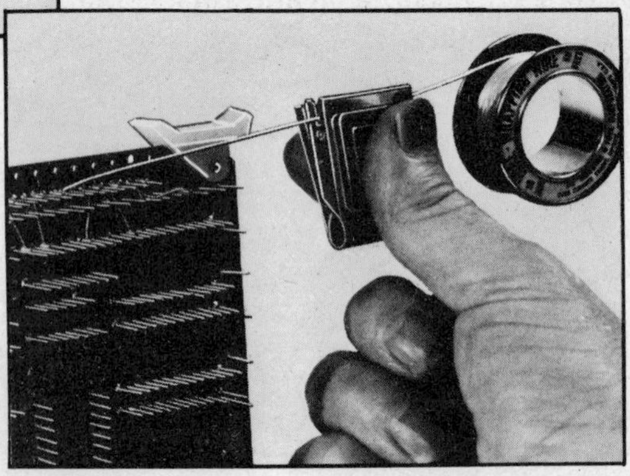

Figure 3-23. A Convenient, Inexpensive Cut-and-Strip Tool

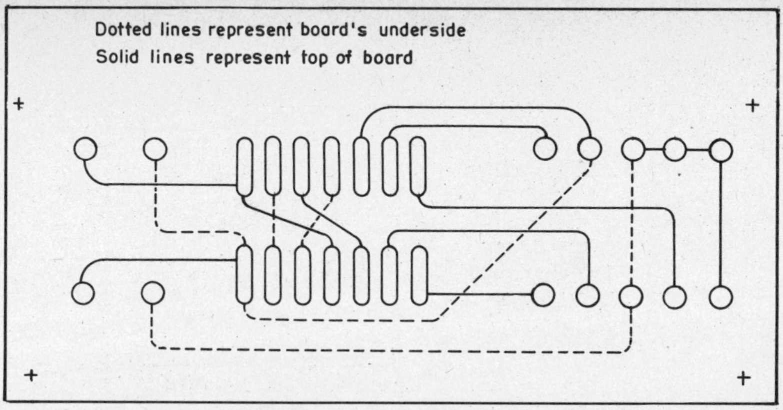


Figure 3-24. A Typical Layout Sketch

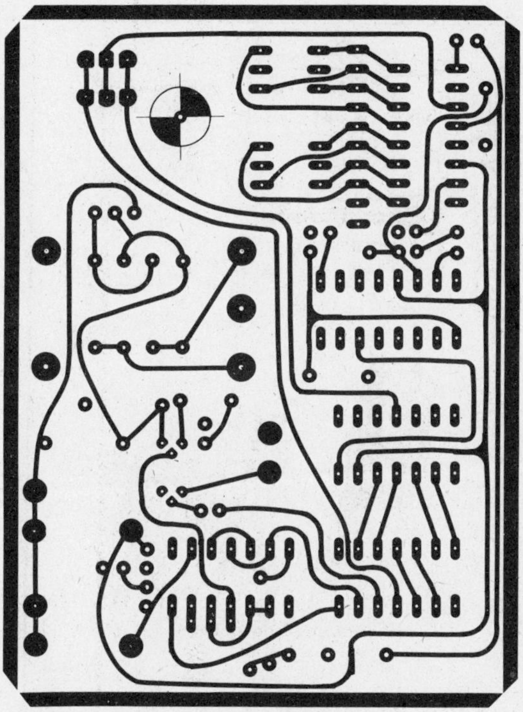

Figure 3-25. A Layout Using Pattern Pre-Cut Artwork

The "cross hair" symbol in the upper left-hand corner is for registration or alignment of the negative on one side of the board with the negative on the other side of the board. Note the presence of the five pre-cut DIP patterns for ICs. The bold lines around the perimeter of the figure are its border. The eventual board has these lines cropped off.

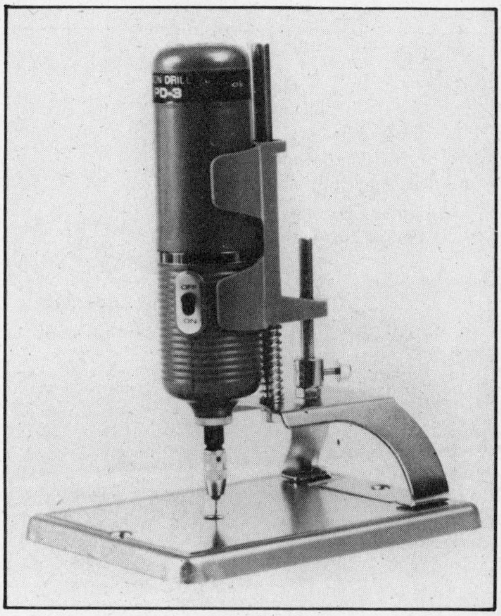


Figure 3-26. A Miniature PC Board Drill

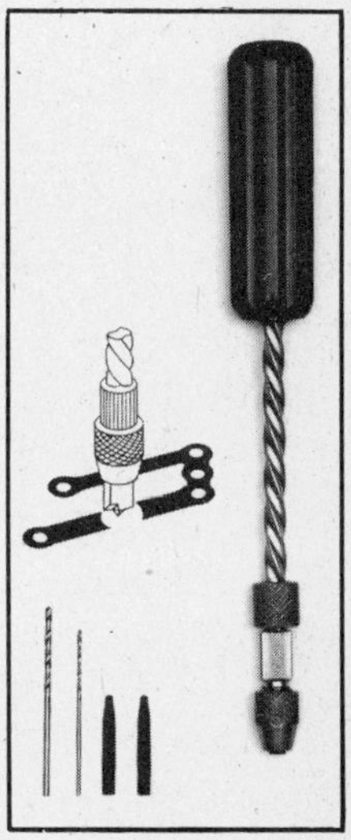

Figure 3-27. A Trace-Cutting Tool

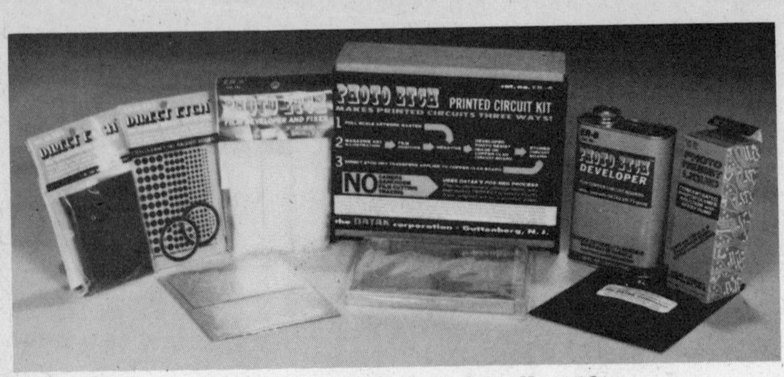


Figure 3-28. A Kit that Photographically Lifts Artwork off a Printed Page

Figure 3-29. Several Types of Solder in Spool Form

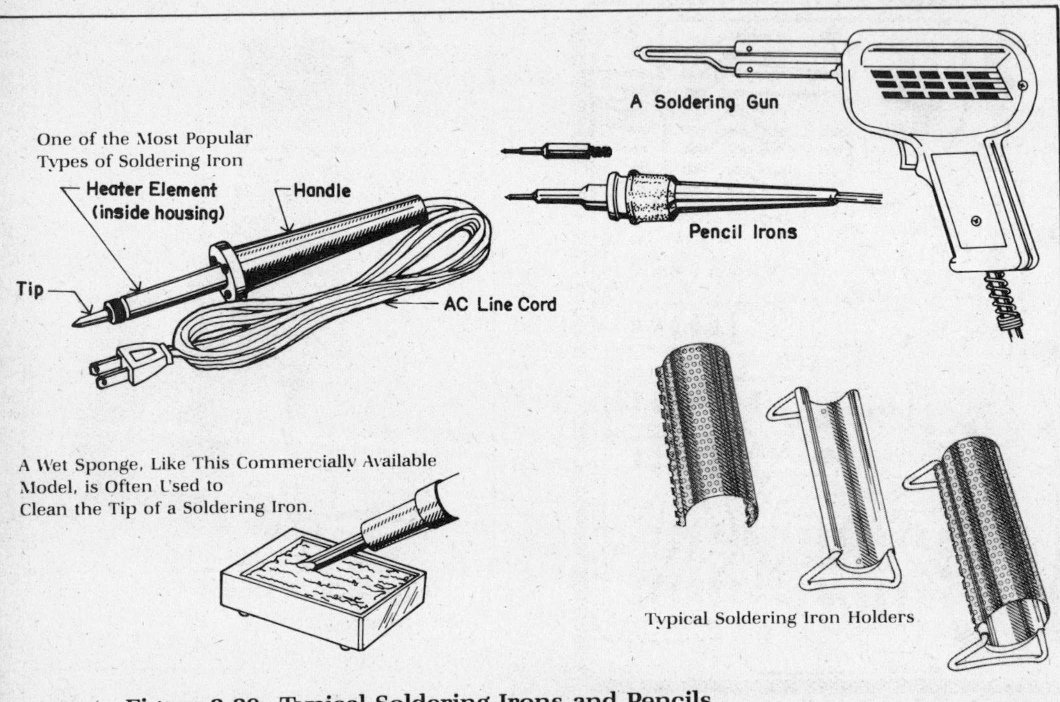


Figure 3-30. Typical Soldering Irons and Pencils

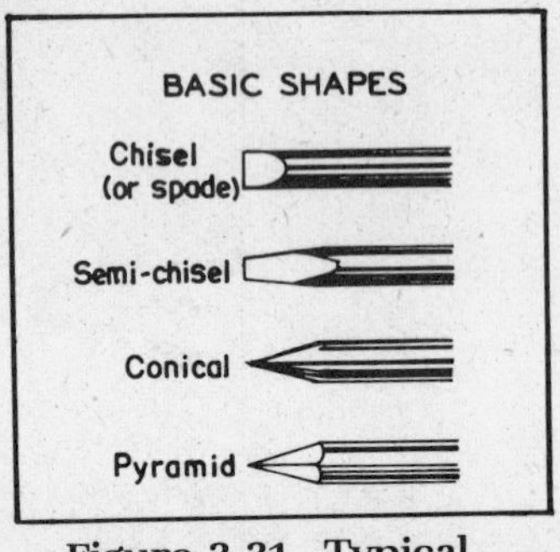


Figure 3-31. Typical Soldering Iron Tips

Figure 3-32. A Molex Pin Socket Soldered onto a PC Board

## Figure 3-33. Illustrated Instructions Showing How to Use Molex Pins (Courtesy of Molex, Inc.)

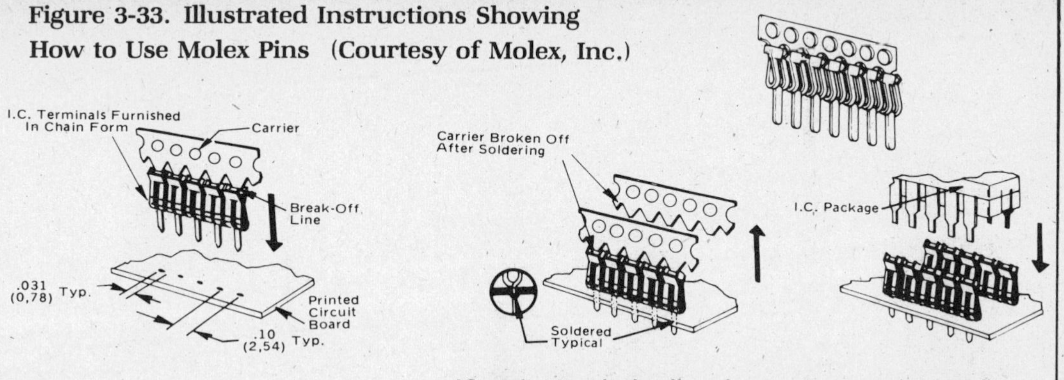


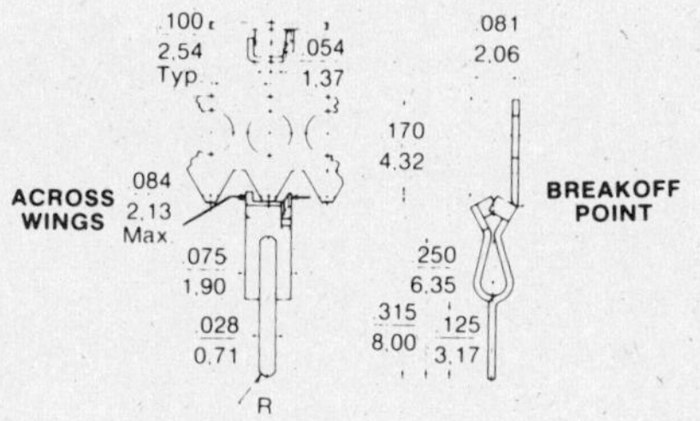


**IC socket terminals offer a low-cost approach to applications specifying plug-in assemblies on the printed circuit board.**

**Method of Assembly:**

1. Terminals are supplied on a carrier in chain form.
2. The printed circuit tails are inserted into the board.
3. After flow-soldering the carrier strip is broken off.
4. The customer's IC package is then inserted into the terminals.

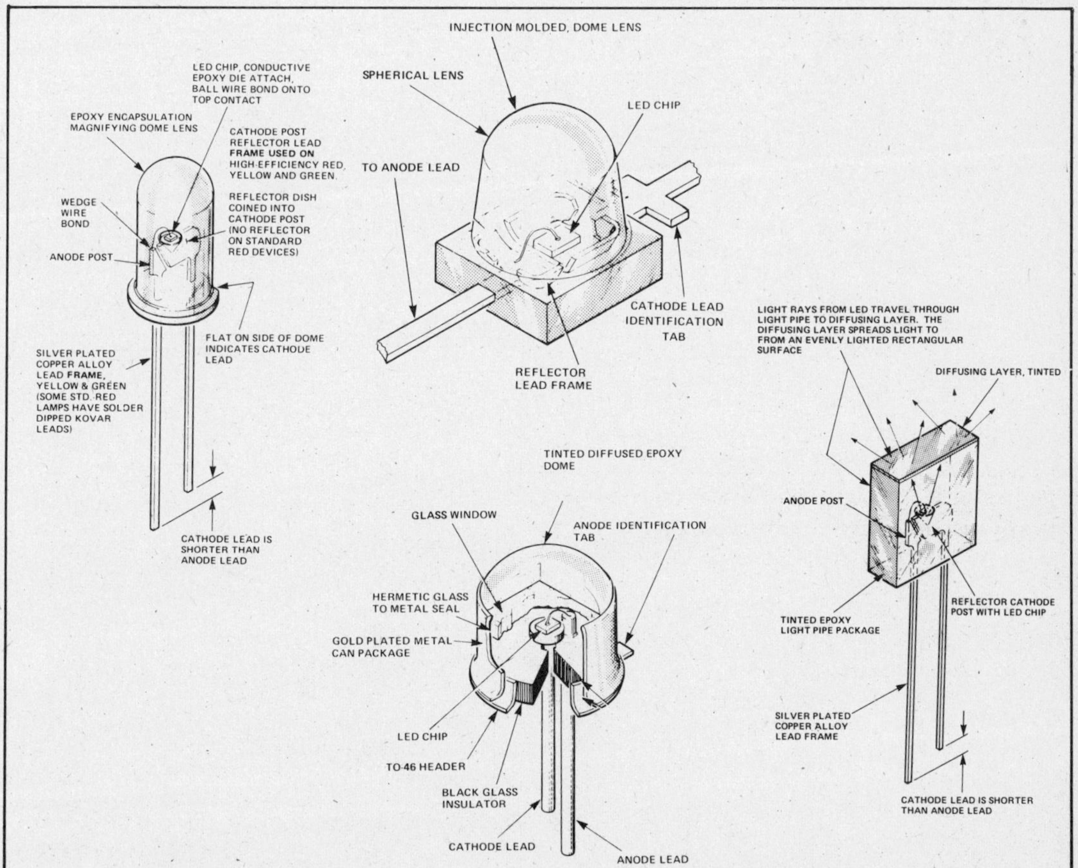


Figure 3-34. A Representative Sampling of Various LED Shapes and Types

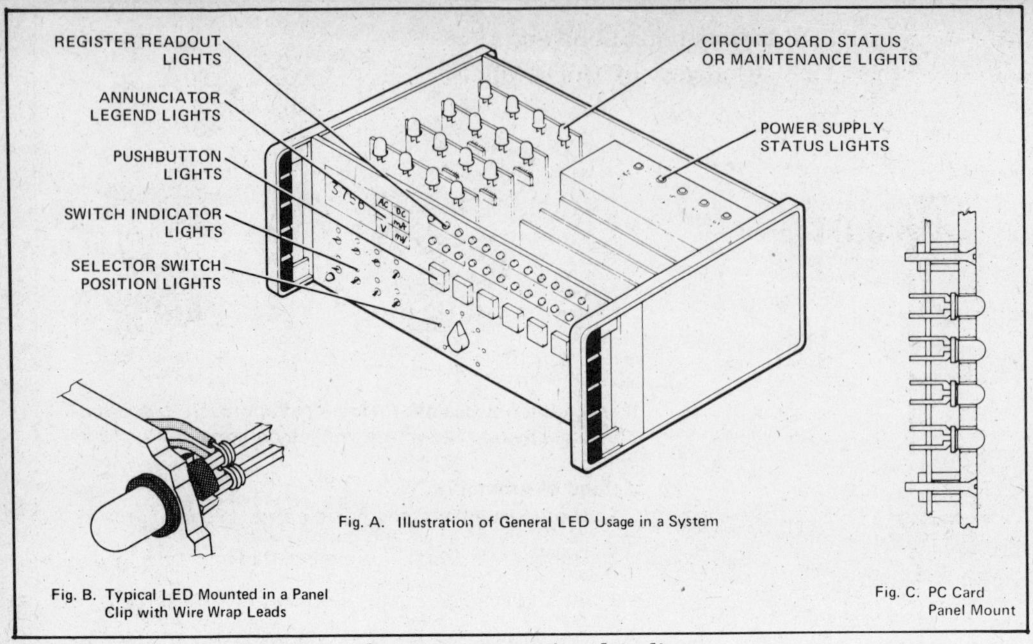


**Figure 3-35. A Typical Use of an LED as a Visual Indicator**

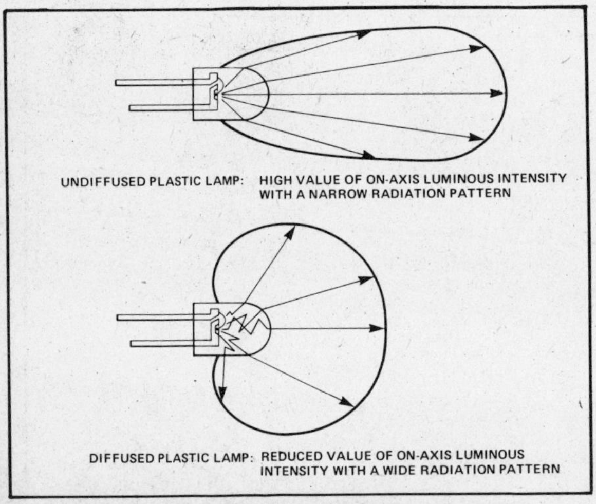


**Figure 3-36. A Light Radiation Pattern Comparison of Different LED Types**

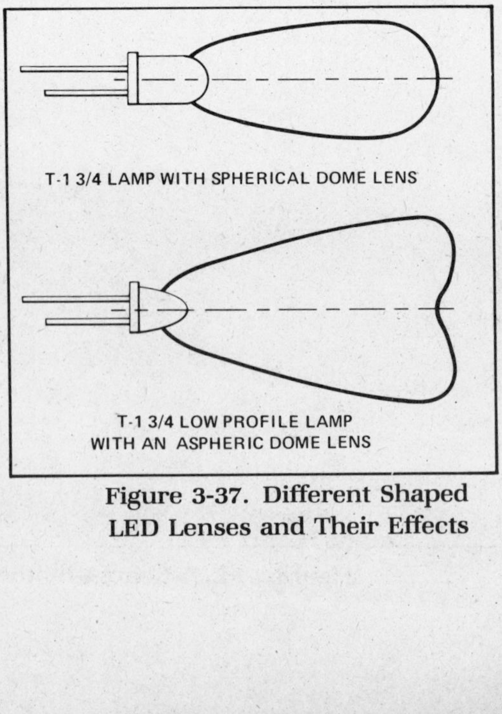


**Figure 3-37. Different Shaped LED Lenses and Their Effects**

Let us briefly review Ohm's Law and its specific application to determining the amount of current that flows through an LED. In the following illustrative example assume that we are working with TTL logic which requires 5 volts DC.

Ohm's Law states that E = IR, or voltage expressed in volts equals the current as expressed amps multiplied by the resistance in ohms. So if we manipulated this simple equation to rearrange the terms we would discover that $I = \frac{E}{R}$ and $R = \frac{E}{I}$. The verification of these two equations can be done by the student. Now, putting this theory to actual use:

Assumptions

1. An assumption is made that a red LED has a 1.8 volt drop across it.
2. The desired current to be drawn is 15 mA; we will discover later that a normal TTL gate can "sink" or receive 16 mA of current flowing into it.
3. The power source is +5V.
4. Current flows from the positive to the negative.

$$R = \frac{E_{SOURCE\ (5V)} - E_{LED\ (1.8V)}}{I(15\text{ mA current})} = \frac{3.2V}{.015\text{ A}} = 213.3\text{ ohms}$$

Use a 220 ohm resistor, the next largest value in standard 5% resistors. The hookup is as follows and will appear again throughout the experiments in this book:

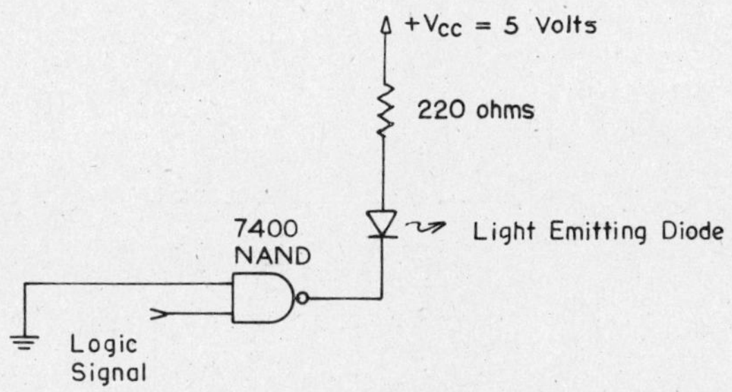


The LED and 220 ohm resistor may be substituted wherever an incandescent lamp appears throughout all experiments in this book. Also, where transistors appear, use a 2N2222 transistor.

# Laboratory Experiments

The experiments presented in this chapter are:

- ☐ The TTL inverter
- ☐ The TTL NAND gate
- ☐ The TTL NOR gate
- ☐ A TTL RS flip-flop using NAND gates
- ☐ A ONE-SHOT multivibrator using TTL NAND gates
- ☐ A multivibrator using TTL NAND gates
- ☐ The CMOS inverter
- ☐ The CMOS NAND gate
- ☐ The CMOS NOR gate
- ☐ The CMOS RS flip-flop using NAND gates
- ☐ A ONE-SHOT multivibrator using CMOS NAND gates
- ☐ An astable multivibrator using CMOS NAND gates in two different configurations plus an astable multivibrator using inverters, Exclusive-OR, and OR gates

These few experiments are very easy and will provide valuable experience with each of these families.

In the following laboratory experiments, three TTL ICs will be used. They are shown here with their pin connections and logic symbols.

The TTL 7400 NAND Gate.

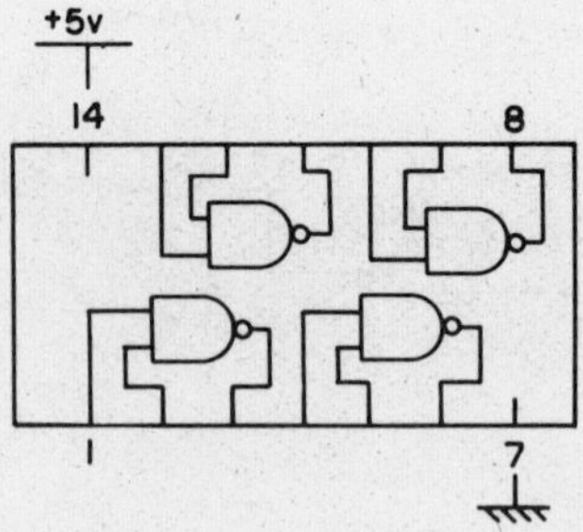

The TTL 7402 NOR Gate.

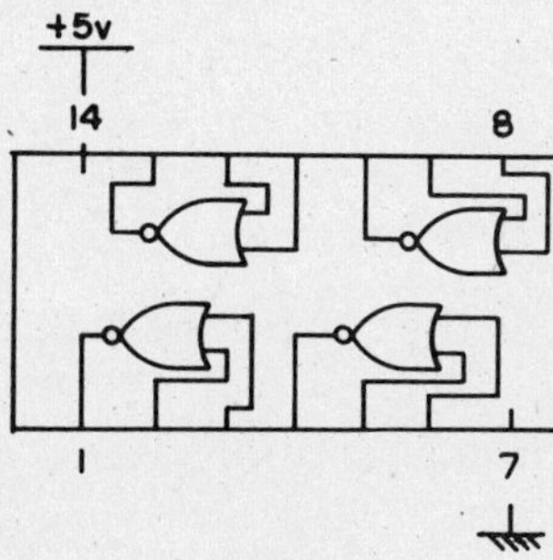


The TTL 7404 Inverter.

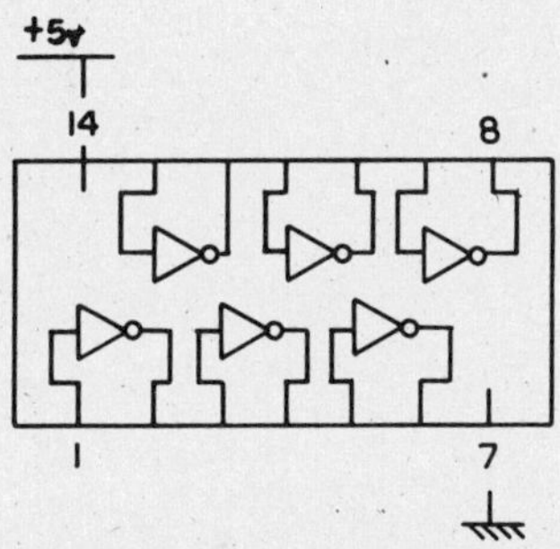


Using the digits 1 and 0 to represent a high and a low voltage is very common in digital electronics.

**Q:** What is used to represent a high voltage?

**A:** The 1.

**Q:** What is used for low voltage?

**A:** The 0.

Using the 1 and the 0 in the above manner is called a "positive logic" system.

**Q:** What is a "positive logic" system?

**A:** This is where a 1 is used to indicate a high voltage level and a 0 is used for a low voltage level.

A "negative logic" system also exists. In this convention the 1 is used for the low voltage and the 0 for the high voltage. In this book though we will only use positive logic.

### *Experiment 1: The 7404 Inverter*

CAUTION: Do not connect the output of any TTL IC to ground.

1. Mount the IC in a breadboard and apply +5V to Pin 14 and ground to Pin 7.

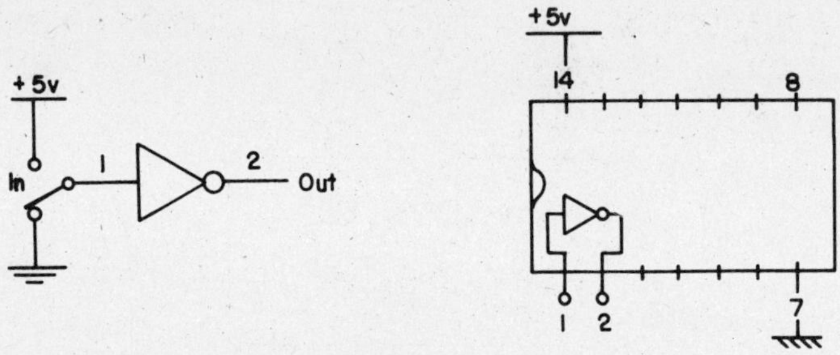


2. Connect the input of Pin 1 to 0V.
3. Measure the output voltage at Pin 2.
4. Enter these two voltages in a truth table. (A truth table is a systematic listing of all possible input combinations to a gate or circuit and the respective resultant outputs.) Use 0 for ground or 0V and 1 for +5V.
5. Connect the input of Pin 1 to +5V.
6. Measure the output voltage at Pin 2.
7. Enter these output results in the same truth table.

**Q:** What function is represented by this truth table?

**A:** Inversion. Your truth table should look like this:

| Pin 1 | Pin 2 |
|---|---|
| 0 | 1 |
| 1 | 0 |

Repeat this for all the other inverters in this IC. You should get the same result.

It is good practice not to leave any logic inputs ungrounded. They might float; that is, the output will be in an intermediate state or not $V_{cc}$ (a 1) or ground (a 0). This intermediate state consumes a lot of power with nothing accomplished.

One extra simple test should be conducted with the TTL inverter.

1. Leave Pin 1 "floating". That is, do not connect it to either ground or +5V.
2. Measure the output as Pin 2. It should be 0V.

With the TTL inverter, this is not an important point. This same experiment is repeated later with the CMOS inverter and both TTL and CMOS NAND and NOR gates—where it is important.

***Experiment 2: The 7400 TTL NAND Gate***

1. Mount the IC in a breadboard and apply +5V to Pin 14 and ground to Pin 7.

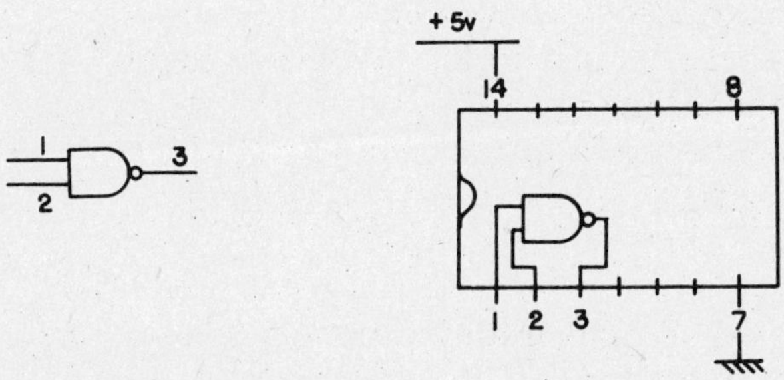


2. Use inputs at Pin 1 and Pin 2. Connect both to ground.
3. Measure the output voltage at Pin 3.
4. Enter all these in a truth table. Use 0 for ground or 0V, and 1 for +5V.
5. Connect Pin 1 to +5V, leave Pin 2 at ground.
6. Measure output voltage at Pin 3.
7. Enter these values in the truth table.
8. Connect Pin 1 to ground and Pin 2 to +5V.
9. Measure the output at Pin 3.
10. Enter these in the truth table.
11. Connect both Pin 1 and Pin 2 to +5V.
12. Measure output voltage at Pin 3.
13. Enter these in the truth table.

**Q:** What function is represented by this truth table?

**A:** This is the NAND function. Your truth table should look like this:

| Pin 1 | Pin 2 | Pin 3 |
|---|---|---|
| 0 | 0 | 1 |
| 0 | 1 | 1 |
| 1 | 0 | 1 |
| 1 | 1 | 0 |

Repeat this experiment for the other sets of inputs and outputs.

### *Experiment 3: The 7400 NAND Gate Used as an Inverter*

1. Use the same IC and setup as in Experiment 2.
2. Connect Pin 2 to +5V.
3. Connect Pin 1 to ground.
4. Measure the output at Pin 3.
5. Enter the Pin 1 and Pin 3 voltages in a truth table.
6. Connect Pin 1 to +5V.
7. Measure the output at Pin 3.
8. Enter these two in the truth table.

**Q:** Draw a simple circuit of the above NAND gate, and draw the truth table for this. What function is this?

**A:** The circuit is:

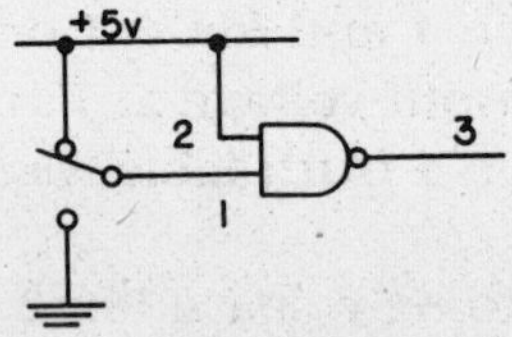


This function is inversion. This is the truth table.

| Pin 1 | Pin 2 | Pin 3 |
|---|---|---|
| 0 | 1 | 1 |
| 1 | 1 | 0 |

(Pin 2 is a 1 at all times.)

**Q:** What simple conclusion can you make from this experiment?

**A:** That a NAND gate can be used as an inverter by connecting one input permanently to a logic 1.

### *Experiment 4: Another 7400 NAND Gate Connection*

1. Use the same setup as in Experiment 3.
2. Connect Pin 1 and Pin 2 together.
3. Connect both to +5V.
4. Measure the output at Pin 3.
5. Enter these in a truth table.
6. Connect both input pins to 0V.
7. Measure the output at Pin 3.
8. Enter these in a truth table.

**Q:** Draw a simple circuit of this, and draw the truth table for this. What function is this?

**A:** The circuit is this.

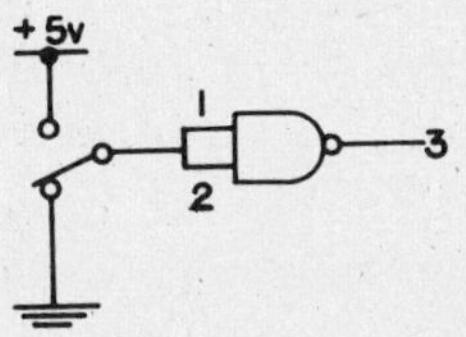


The function is inversion. The truth table is this.

| Pin 1 | Pin 2 | Pin 3 |
|---|---|---|
| 0 | 0 | 1 |
| 1 | 1 | 0 |

### *Experiment 5: The Inhibited NAND Gate*

1. Use the same NAND gate setup.
2. Connect Pin 2 to 0V.
3. Connect Pin 1 to 0V.
4. Measure the output voltage at Pin 3.

5. Enter the voltages for Pin 1 and Pin 3 in a truth table.
6. Connect Pin 1 to +5V.
7. Measure the voltage at Pin 3.
8. Enter these values in the truth table.

**Q:** Draw a simple circuit of this, and draw the truth table for this. What is happening here? (Hint: It is not an inversion.)

**A:** This is the circuit.

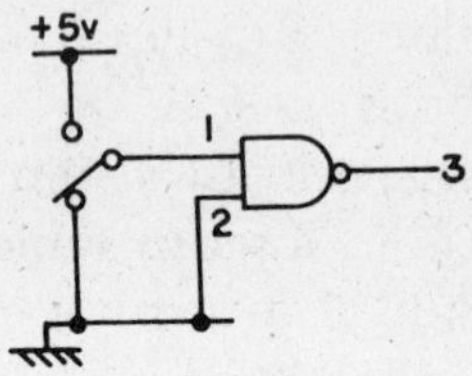


This is the truth table.

| Pin 1 | Pin 2 | Pin 3 |
|---|---|---|
| 0 | 0 | 1 |
| 1 | 0 | 1 |

(Pin 2 is a 0 at all times.)

The permanent 0 on Pin 2 has kept the output on Pin 3 at 1. The NAND gate had been "inhibited" from working by the "inhibit" placed on the input at Pin 2. A logical 0 inhibits a NAND gate.

**Q:** When a NAND gate is inhibited, what is its output?

**A:** It is always a 1.

### *Experiment 6: The 7402 TTL NOR Gate*

1. Mount this IC in a breadboard and apply +5V to Pin 14 and ground to Pin 7.

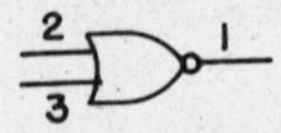

2. Use inputs at Pin 2 and Pin 3. Connect both to ground.
3. Measure the output voltage at Pin 1.
4. Enter all these values in a truth table.
5. Connect Pin 2 to +5V. Leave Pin 3 at OV.
6. Measure the output at Pin 1.
7. Enter these values in the truth table.
8. Connect Pin 2 to OV and Pin 3 to +5V.
9. Measure the output at Pin 1.
10. Enter these in the truth table.
11. Now connect both inputs to +5V.
12. Measure the output.
13. Enter these in the truth table.

**Q:** What function is represented by this truth table?

**A:** This is the truth table for a NOR.

| Pin 2 | Pin 3 | Pin 1 |
|---|---|---|
| 0 | 0 | 1 |
| 0 | 1 | 0 |
| 1 | 0 | 0 |
| 1 | 1 | 0 |

### *Experiment 7: The 7402 NOR Gate Used as an Inverter*

1. Keep the setup of Experiment 6.
2. Connect Pin 2 to ground.
3. Connect Pin 3 to ground.
4. Measure the output at Pin 1.
5. Enter these values in a truth table.
6. Connect Pin 3 to +5V.
7. Measure the output at Pin 1.
8. Enter these in the truth table.

**Q:** Draw a simple circuit of this, and draw the truth table for this. What function is this?

**A:** This is the circuit.

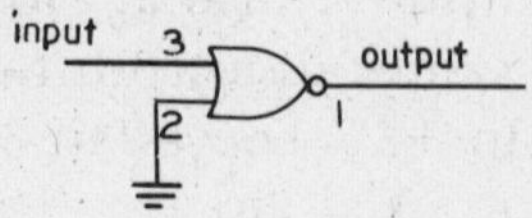


This function is inversion. This is the truth table.

| Pin 3 | Pin 2 | Pin 1 |
|---|---|---|
| 0 | 0 | 1 |
| 1 | 0 | 0 |

(Pin 2 is a 0 at all times.)

***Experiment 8: Another 7402 NOR Gate Connection***

1. Use the same setup as previously used.
2. Connect Pins 2 and 3 together.
3. Connect both to +5V.
4. Measure the output at Pin 1.
5. Enter these values in a truth table.
6. Connect both pins to OV.
7. Measure the output voltage at Pin 1.
8. Enter these in the truth table.

**Q:** Draw the circuit of this. Draw the truth table. What function is this?

**A:** This is the circuit.

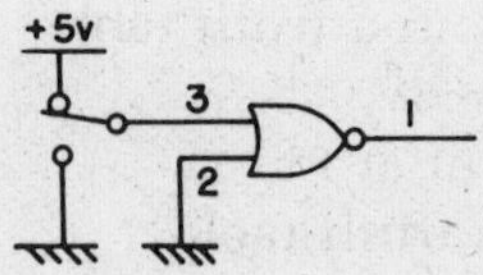


This function is inversion. This is the truth table.

| Pin 2 | Pin 3 | Pin 1 |
|---|---|---|
| 0 | 0 | 1 |
| 1 | 1 | 0 |

When all inputs to a NOR or a NAND gate are connected together, the output will be the inversion of the input.

### *Experiment 9: The Inhibited TTL 7402 NOR Gate*

1. Continue to use the setup of Experiment 6.
2. Connect Pin 2 to +5V.
3. Connect Pin 3 to OV.
4. Measure the output at Pin 1.
5. Enter these in a truth table.
6. Connect Pin 3 to +5V.
7. Measure the output at Pin 1.
8. Enter these in the truth table.

**Q:** Draw the circuit. Draw the truth table. What is happening to the output?

**A:** This is the circuit.

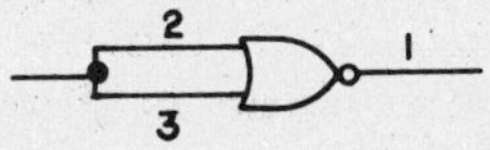


This is the truth table.

| Pin 3 | Pin 2 | Pin 1 |
|---|---|---|
| 0 | 1 | 0 |
| 1 | 1 | 0 |

(Pin 2 is a 1 at all times.)

The permanent logic 1 on Pin 2 has kept the output on Pin 1 at logic 0. A logic 1 on an input inhibits the gate with its output at 0.

**Q:** When the NOR gate is inhibited, what is its output?

**A:** Always 0.

### *Experiment 10: The RS Flip-Flop Using TTL NAND Gates*

1. Set up this circuit.

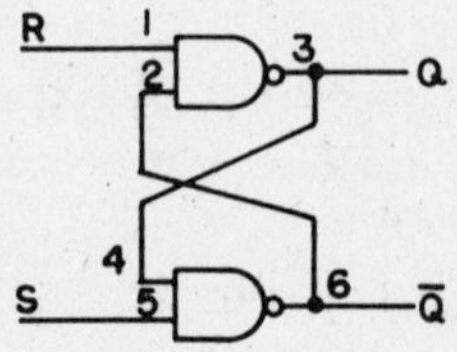


2. Set Pin 1 to +5V.
3. Set Pin 5 to OV.
4. Measure the outputs at Q and $\overline{Q}$.
5. Enter these in a truth table.
6. Keep Pin 1 at +5V, put Pin 5 to OV.
7. Measure the outputs.
8. Enter these values in the truth table.
9. Keep Pin 1 at +5V, return Pin 5 to OV.
10. Measure the outputs.
11. Enter these in the truth table. The three lines of the truth table should be the same as shown here.

| R | S | Q | $\overline{Q}$ |
|---|---|---|---|
| 1 | 0 | 0 | 1 |
| 1 | 1 | 0 | 1 |
| 1 | 0 | 0 | 1 |

From this we can see that keeping one input at a logic 1 inhibits the outputs.

12. Keep Pin 1 at +5V and Pin 5 at OV.
13. Measure the outputs.
14. Enter these in a new truth table.
15. Set Pin 1 to OV, keep Pin 5 at OV.
16. Measure the outputs.
17. Enter these in the truth table.
18. Return Pin 1 to +5V, keep Pin 5 at 0V.
19. Measure the outputs.
20. Enter these in the truth table. The three lines of the table should be the same as shown on the following page.

| R | S | Q | $\overline{Q}$ |
|---|---|---|---|
| 1 | 0 | 0 | 1 |
| 0 | 0 | 1 | 1 |
| 1 | 0 | 0 | 1 |

This shows that keeping one input at a logic 0 while the other input varies will produce the "not allowed" input condition of both $\overline{Q}$ and Q at a logic 1. In practice, this situation is to be avoided.

21. Set Pin 1 to +5V and Pin 5 to OV.
22. Measure the outputs.
23. Enter these in a new truth table.
24. Set Pin 1 to OV, and simultaneously set Pin 5 to +5V.
25. Measure the outputs.
26. Enter these in the table.
27. Simultaneously return Pin 1 to +5V and Pin 5 to OV.
28. Measure the outputs.
29. Enter these in the table. The three lines of the table are these.

| R | S | Q | $\overline{Q}$ |
|---|---|---|---|
| 1 | 0 | 0 | 1 |
| 0 | 1 | 1 | 0 |
| 1 | 0 | 0 | 1 |

Here the outputs change state as the inputs change state. The simultaneous changing of the inputs is needed for the correct operation of the "RS" Flip-Flop.

**Q:** What two useful practical points have you just learned about the NAND gate RS Flip-Flop?

**A:** First: a logic 1 on an input will inhibit it.
Second: that both inputs must change state at the same time for correct operation.

***Experiment 11: The TTL NAND Gate ONE-SHOT (Monostable Multivibrator)***

1. Set this circuit up, then apply power.

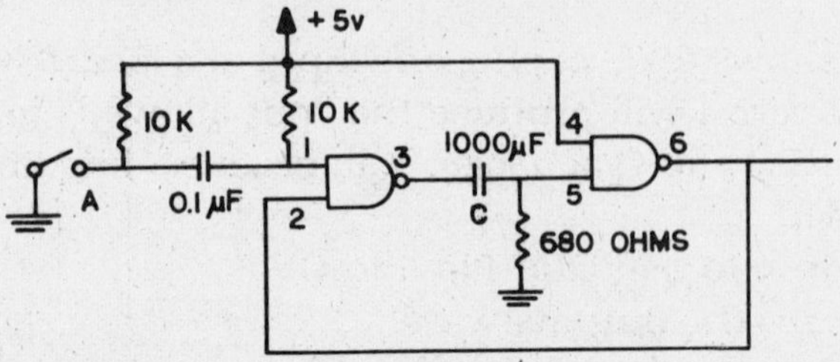


2. Set the oscilloscope vertical input to 1V/cm. Center the trace on the screen.
3. Set the horizontal timing to 0.5 cms/sec.
4. Measure the voltages at each pin with the oscilloscope. They should be as shown in this table.

| Pin Number | Voltage (Logic states) |
|---|---|
| Point A | 5V |
| 1 | 5 |
| 2 | 5 |
| 3 | 0 |
| 4 | 5 |
| 5 | 0 |
| 6 | 5 |

Note: With TTL a high can actually be anywhere between 3.5V and 5V. A TTL low can be anywhere between 0V and 0.3V.

These same voltages are also shown at time $T_1$ in the waveform drawings.

5. Close the switch and hold it closed.
6. Observe the voltage changes at each pin. (Adjust the oscilloscope "trigger" control to produce a stationary waveform. You may need help with this.)
7. Repeat steps 5 and 6 as many times as needed to make all the observations.
8. Draw all the observed waveforms. Show all the times and the voltage levels. These should be the same as shown opposite.

9. Release the switch so that it is open.
10. Observe point A with one probe and Pin 6 with the other.
11. Momentarily close the switch, and allow it to immediately open.
12. Draw the waveforms observed. They are shown here. The duration of the negative-going output pulse at Pin 6 can be calculated from:

$$T_D = 1.2 \times R \times C$$

R is expressed in ohms and C in Farads.

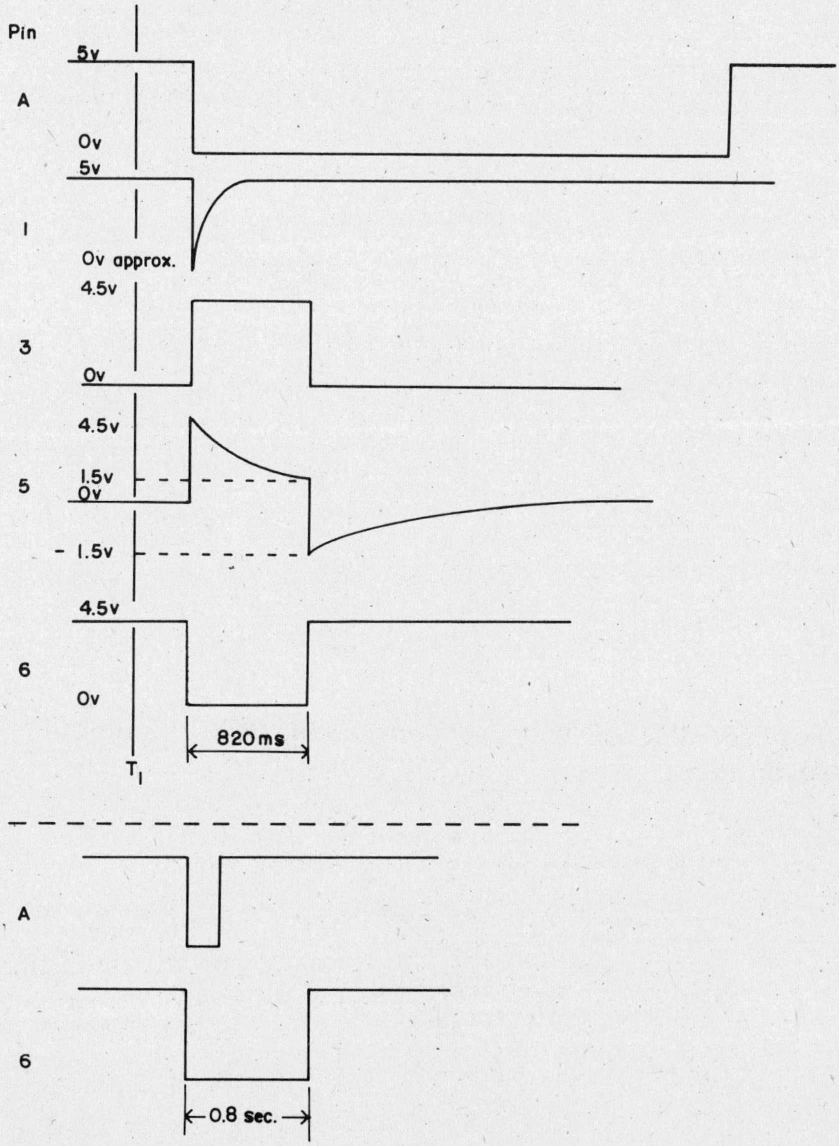

**Q:** Calculate the duration of the pulse in the above experiment.

**A:** This is the calculation:

$$T_D = 1.2 \times 680 \times 1000 \times 10^{-6}$$
$$= 816 \text{ millisecs}$$

From the circuit, C = 1000 uF, R = 680 ohms.

**Q:** Compare this with the observed pulse duration. How close are they?

**A:** The observed pulse duration in the experiment will vary somewhat. This is because capacitors are NOT ACCURATE in their values. They can be 50% low to as much as 100% high. Times between 500 ms and 1300 ms should be expected.

### *Experiment 12*

Repeat Experiment 11 with different values for capacitor C. Measure the output pulse duration for each capacitor. See if you get values close to those in this table.

| C | $T_D$ |
|---|---|
| 20 uF | 16 ms |
| 50 | 41 |
| 100 | 82 |
| 200 | 160 |
| 500 | 410 |
| 1000 | 820 |

### *Experiment 13: The TTL NAND Gate Astable (Free-Running) Multivibrator*

1. Set this circuit up and apply power.

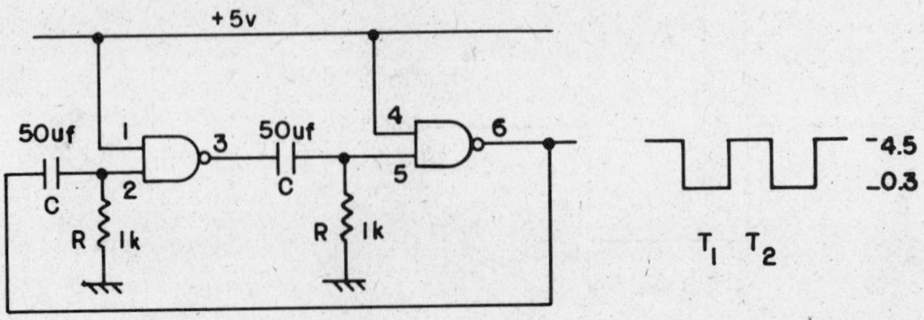

2. Set the oscilloscope vertical input to 1V/cm. Center the trace on the screen.
3. Set the horizontal timing to 10 msec/cm.
4. Observe the voltage waveforms at Pin 6, Pin 2, Pin 3, and Pin 5.
5. Measure the voltage levels and the time durations of the pulses.
6. Draw waveform diagrams showing these voltages and times. They are shown here. Pins 1 and/or 4 could be tied to a logic high as well as the +5V supply. This would allow the oscillator to be shut off or turned on electronically.

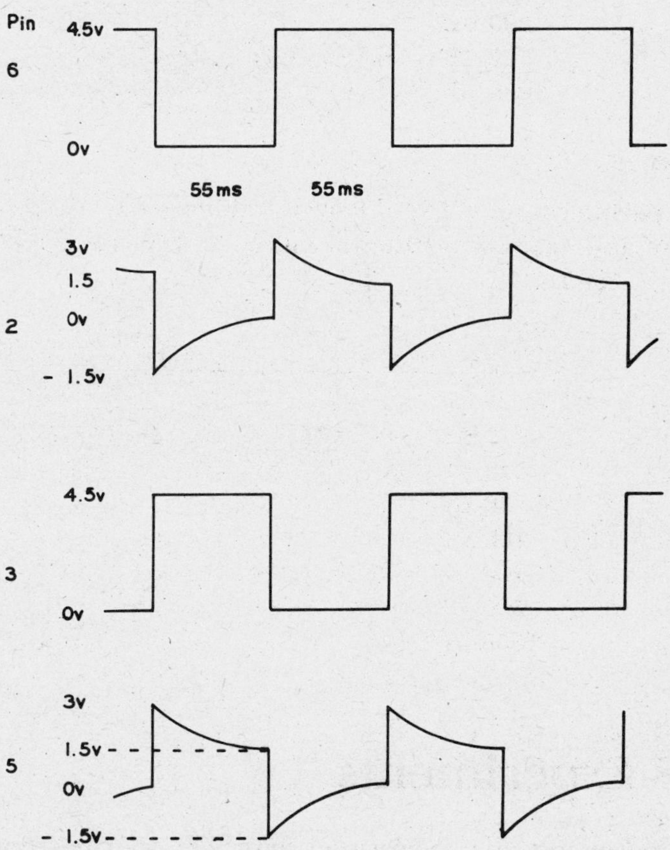


NOTE:

- ☐ Exact voltage levels will vary from gate to gate
- ☐ 0V can be anywhere from 0V to 0.3V
- ☐ $T_1$ and $T_2$ may be slightly different due to component tolerances

***Experiment 14***

Repeat Experiment 13 using different values for $C_1$ and $C_2$. Enter the values of $T_1$ and $T_2$ that you measure with the oscilloscope in this table.

| $C_1 = C_2$ | $T_1$ | $T_2$ |
|---|---|---|
| 1 uF | | |
| 5 uF | | |
| 10 uF | | |
| 50 uF | | |
| 100 uF | | |
| 200 uF | | |
| 500 uF | | |
| 1000 uF | | |

***Experiment 15***

Repeat Experiment 13 using different values for $C_2$. Use the values shown here and again enter the measured times of $T_1$ and $T_2$.

Keep $C_1$ = 200 uF.

| $C_2$ | $T_1$ | $T_2$ |
|---|---|---|
| 1 uF | | |
| 5 uF | | |
| 10 uF | | |
| 50 uF | | |
| 100 uF | | |
| 200 uF | | |
| 500 uF | | |
| 1000 uF | | |

# CMOS Experiments

All the experiments just performed with TTL can be repeated with CMOS digital ICs, as this family contains the same simple logic functions.

However, three things are different:

1. CMOS will work on any supply voltage between +3V and +15V or +20VDC if "B" suffix CMOS is used.

2. The functions are connected to pins which are different from those in TTL. The diagrams will show this.
3. The CMOS internal circuitry is very different.

What these three points mean is very simple—you cannot just pull out a TTL from a breadboard and insert a CMOS in its place. The connections to the pins must be rewired.

Each of the experiments conducted with the TTL ICs is to be repeated using the CMOS ICs. The steps are exactly as before and so they will not be discussed again. Just refer back to the procedure written for the TTL.

In each case, the pin connections for the CMOS ICs and the circuit are shown.

Use a 9V transistor radio battery to power the CMOS. CMOS consumes much less power than TTL does!

**Q:** Why not just pull out the TTL IC and plug in a CMOS?

**A:** The pin connections are different.

**Q:** What must be done because of this?

**A:** Rewire the breadboard circuit.

In the following CMOS experiments, three CMOS ICs will be used. They are shown here with their pin connections and logic symbols.

The CMOS 4011 NAND Gate.

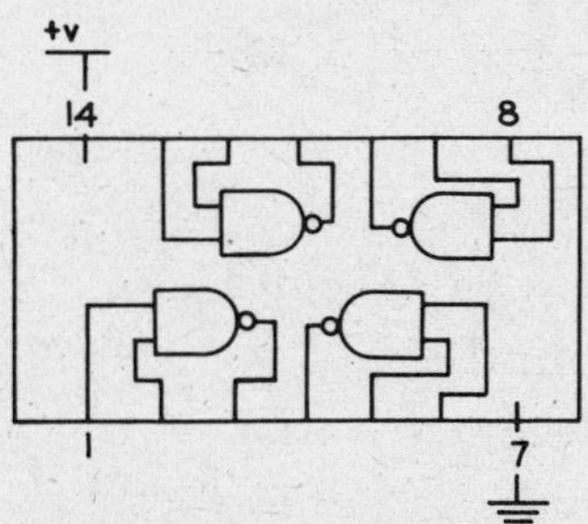

The CMOS 4001 NOR Gate.

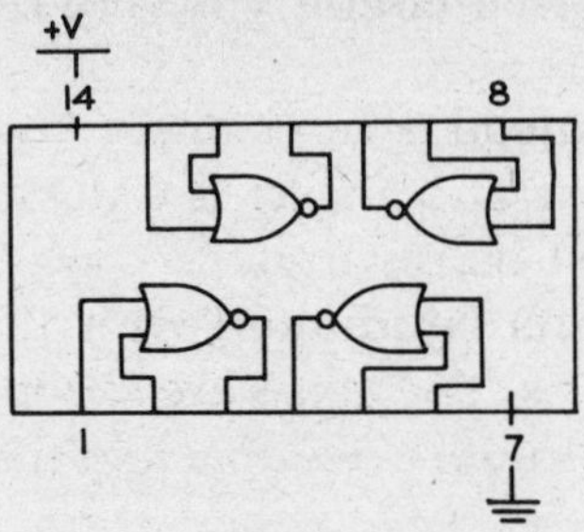


The CMOS 4049 Inverter.

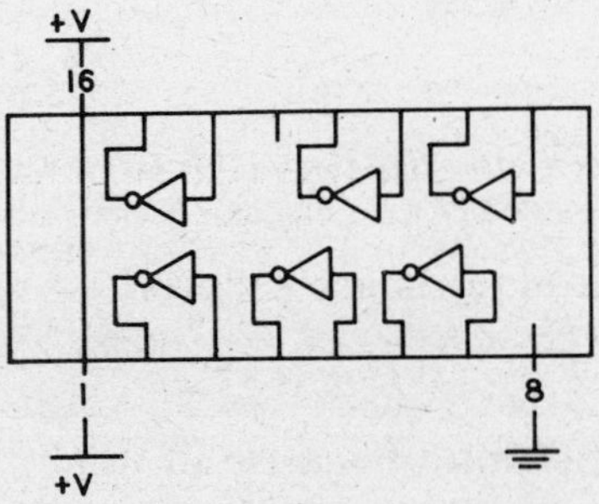


### *Experiment 1: The CMOS 4049 Inverter*

Repeat the TTL inverter Experiment 1 using the CMOS 4049 IC. The circuit is this.

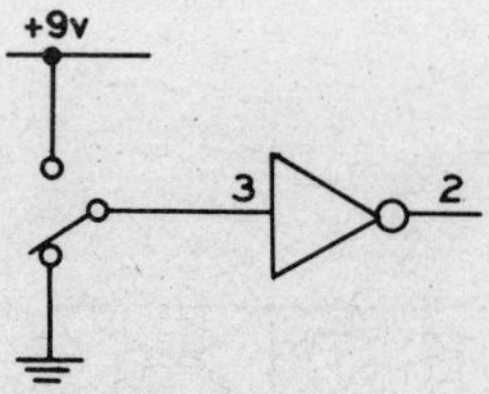


1. Connect Pin 1 and Pin 16 to +9V.
2. Connect Pin 8 to ground.
3. Pin 13 is unused.
4. Leave Pin 3 "floating"—that is—do not connect it to anything.

This is the truth table.

| Pin 3 | Pin 2 |
|---|---|
| 0 | 1 |
| 1 | 0 |

**Q:** What is the output at Pin 2?

**A:** It is +9V, a logic 1.

This is the usual case when an inverter input in CMOS is allowed to float. However, often it will pick up a 60 Hz signal from fluorescent lights and other equipment, and the output will be very unstable. In general, do not let inputs float—use them!

***Experiment 2: The CMOS 4011 NAND Gate***

Repeat TTL Experiment 2 with this CMOS IC. The circuit is this.

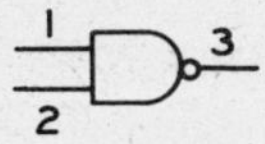


1. Connect Pin 14 to +9V.
2. Connect Pin 7 to ground.
3. Now allow Pin 2 to float.
4. Put Pin 1 to 0V.

The truth table is this.

| Pin 1 | Pin 2 | Pin 3 |
|---|---|---|
| 0 | 0 | 1 |
| 0 | 1 | 1 |
| 1 | 0 | 1 |
| 1 | 1 | 0 |

**Q:** What is the output at Pin 3?

**A:** It is +9V, a logic 1.

With Pin 2 still floating, put Pin 1 to +9V.

**Q:** What is the output at Pin 3?

**A:** It is still at +9V, a logic 1.

Now allow both Pin 1 and Pin 2 to float.

**Q:** What is the output at Pin 3?

**A:** It remains at a logic 1 or +9V.

So if any one input to a CMOS NAND gate is allowed to float, the output is inhibited at a logic 1. This shows that the inputs to CMOS NAND gates should never be allowed to float.

### *Experiment 3: The CMOS NAND Gate Used as an Inverter*

Repeat TTL Experiment 3 with the CMOS NAND gate IC.
The circuit is this.

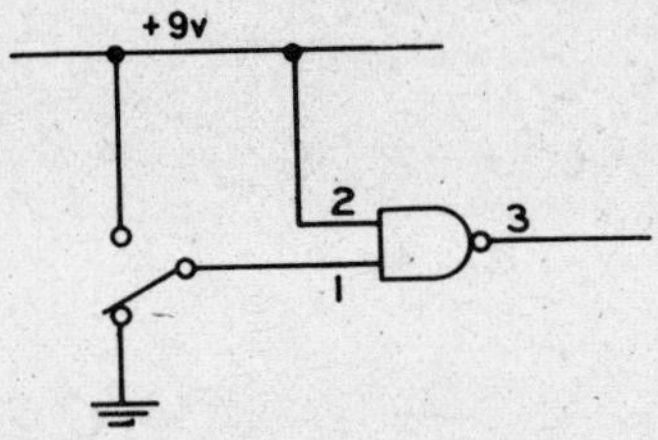


The truth table is this.

| Pin 1 | Pin 3 |
|---|---|
| 0 | 1 |
| 1 | 0 |

Pin 2 is at 1 all the time. This is inversion.
You should now be able to answer this question.

**Q:** What should you do with unused inputs on this IC?

**A:** Connect them to a logic 1.

### *Experiment 4: The CMOS NAND Inverter*

Repeat Experiment 4 of the TTL series. This is the circuit.

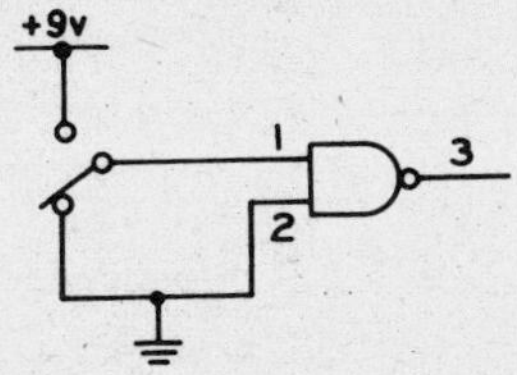


This is inversion. This is the truth table.

| Pin 1 | Pin 2 | Pin 3 |
|---|---|---|
| 0 | 0 | 1 |
| 1 | 1 | 0 |

### *Experiment 5: The Inhibited CMOS 4011 NAND Gate*

Repeat the TTL Experiment 4. This is the circuit.

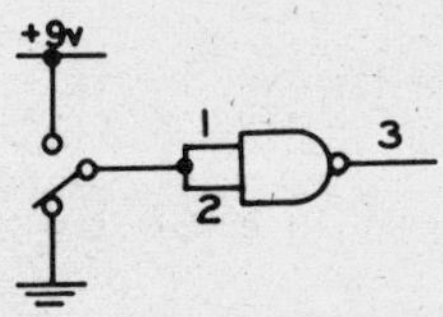


This is the truth table.

| Pin 1 | Pin 2 | Pin 3 |
|---|---|---|
| 0 | 0 | 1 |
| 1 | 0 | 1 |

(Pin 2 is a 0 at all times.)

This shows that a 0 on any input will inhibit the output at a logic 1.

### *Experiment 6: The 4001 CMOS NOR Gate*

Repeat the TTL Experiment 6 with this CMOS IC. The circuit is this.

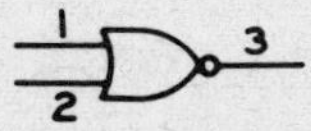

1. Let Pin 1 float.
2. Set Pin 2 to +9V, a logic 1.

This is the truth table.

| Pin 1 | Pin 2 | Pin 3 |
|---|---|---|
| 0 | 0 | 1 |
| 0 | 1 | 0 |
| 1 | 0 | 0 |
| 1 | 1 | 0 |

**Q:** What is the output?

**A:** It should be at ground, or logic 0.

Now with Pin 1 floating, set Pin 2 to ground, a logic 0.

**Q:** What is the output now?

**A:** It should be at +9V, a logic 1.

With a floating input, the CMOS NOR gate appears to act like an inverter; however, a floating input can still pick up noise, so, in general, do not let inputs float.

### *Experiment 7: The CMOS NOR Used as an Inverter*

This is a repeat of TTL Experiment 7. The circuit is this.

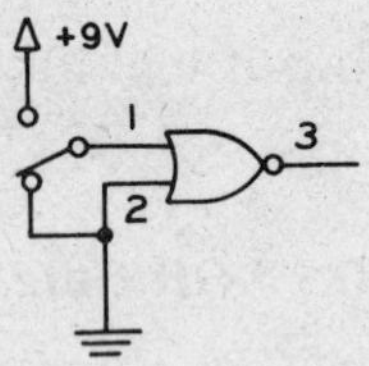


This is an inverter. The truth table is this.

| Pin 1 | Pin 2 | Pin 3 |
|---|---|---|
| 0 | 0 | 1 |
| 1 | 0 | 0 |

(Pin 2 is a logic 0.)

### *Experiment 8: An Alternative CMOS NOR Inverter*

This is TTL Experiment 8 with the CMOS IC. The circuit is this.

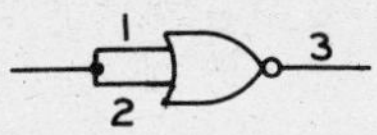


The truth table is this.

| Pin 1 | Pin 2 | Pin 3 |
|---|---|---|
| 0 | 0 | 1 |
| 1 | 1 | 0 |

### *Experiment 9: The Inhibited CMOS NOR Gate*

This is the same as the TTL inhibited NOR gate. This is the circuit.

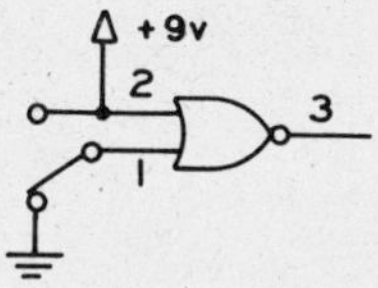


This is the truth table.

| Pin 1 | Pin 2 | Pin 3 |
|---|---|---|
| 0 | 1 | 0 |
| 1 | 1 | 0 |

(Pin 2 is a 1 at all times.)

### *Experiment 10: An RS Flip-Flop with CMOS NAND Gates*

This is a repeat of the TTL RS NAND Flip-Flop. It should be worked through—step by step—just as the TTL experiment was. The truth table and the conditions are the same.

This is the circuit.

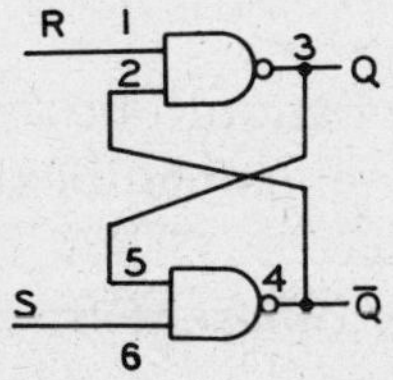

### *Experiment 11: The CMOS NAND "ONE-SHOT"*

This is the same as the TTL NAND "ONE-SHOT" experiment. Use this circuit with the values shown here.

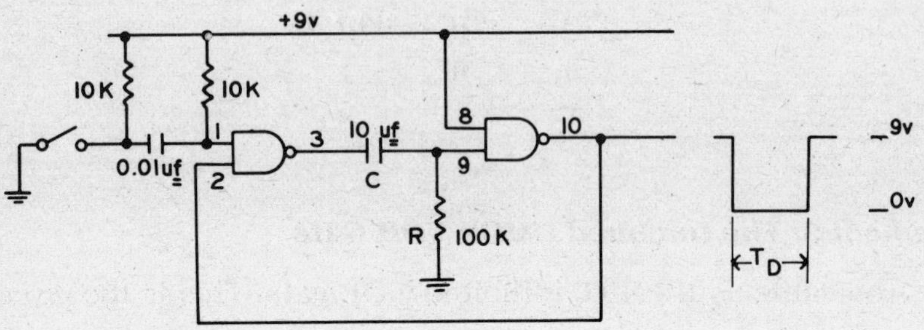


Draw the waveforms at Pins 3, 9, and 10. They will be similar to the TTL case, but not quite the same.

Measure the time duration of the output pulse. Calculate it from this formula:

$$T_D = R \times C$$

This approximate formula gives fairly close results.

**Q:** Do the calculations agree with the observed values?

**A:** Calculated value for the time duration:

$$\begin{aligned} T_D &= R \times C \\ &= 100\text{k} \times 10\ \text{uF} \\ &= 100 \times 10^3 \times 10 \times 10^{-6} \\ &= 1000\ \text{msecs} \\ &= 1\ \text{second} \end{aligned}$$

An observed value should be fairly close to this. Component tolerances are responsible for the differences observed.

The reason this formula is different from the TTL case is due to the internal circuit differences between TTL and CMOS.

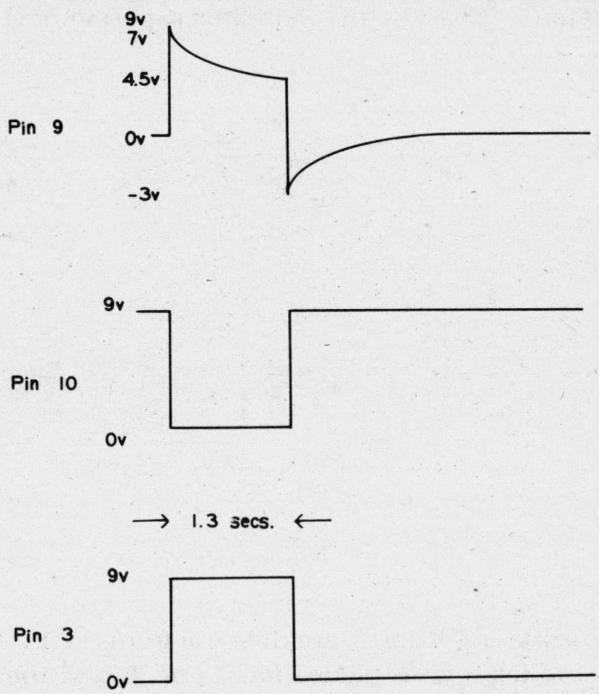


### *Experiment 12*

Repeat Experiment 11 with the values of C and R shown in this table. See how closely your calculated and observed figures agree.

| C | R | $T_D$ Calculated | $T_D$ Observed |
|---|---|---|---|
| 0.1 uF | 100k | | |
| 1 uF | | | |
| 10 | | | |
| 50 | | | |
| 0.1 | 10k | | |
| 1 | | | |
| 10 | | | |
| 50 | | | |
| 100 | | | |

### *Experiment 13: The CMOS NAND Multivibrator*

This is the same as the TTL multivibrator experiment. The circuit for the CMOS IC is shown here.

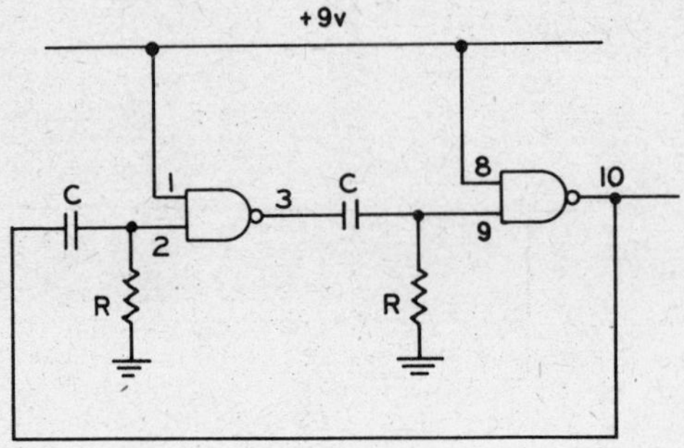


Note that the values for R and C are different from the TTL experiment.

The following table gives values for C and R and the times for the squarewave output duration. Use these values and see if you get similar times.

Draw waveforms at Pins 10, 2, 3, and 9.

**Table 1.** Use R = 100k

| C | $T_1$ | $T_2$ |
|---|---|---|
| 0.002 uF | 250 usec | 200 usec |
| 0.005 | 490 | 490 |
| 0.01 | 800 | 820 |
| 0.033 | 3.3 ms | 3.3 ms |
| 0.1 | 6.4 ms | 6.2 |

**Table 2.** Use R = 10k

| C | $T_1$ | $T_2$ |
|---|---|---|
| 0.1 uF | 600 us | 600 us |
| 1 uF | 8 ms | 7 ms |
| 10 | 100 ms | 100 ms |

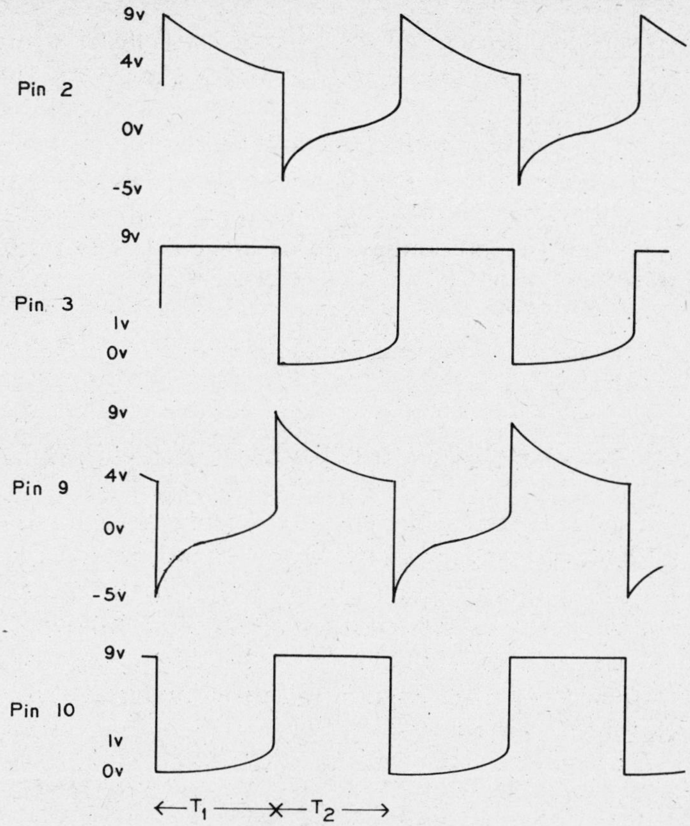


### *Experiment 14: The Asymmetric CMOS NAND Multivibrator*

Repeat the previous experiment using the values for $C_1$ and $C_2$ shown in this table.

Draw the output waveform, and measure the pulse durations. Record them in the table.

| $C_1$ | $C_2$ | $T_1$ | $T_2$ |
|---|---|---|---|
| 1 uF | 0.1 uF | 6.5 ms | 64 ms |
| 0.01 | 0.1 | 6.2 ms | 0.9 ms |
| 0.033 | 0.01 | 3.1 | 0.9 |
| 0.033 | 0.005 | 3.0 | 0.5 |

The waveform will look like this: or this:

Choose two similar values for $C_1$ and $C_2$ and use different values for $R_1$ and $R_2$. Enter the times in a table and draw the waveform. The results will be similar to the above.

Free-running squarewave oscillators or astable multivibrators may be constructed from CMOS inverters, NAND gates, NOR gates, or Exclusive-OR gates. The inverter, NAND, and NOR gates are all actually very similar because they are all configured as inverters. The Exclusive-OR gate is different.

**Q:** Which astable multivibrator gate is distinctively different from the other?

**A:** The Exclusive-OR gate oscillator.

TTL gates (the 7400 series) have a +5V supply voltage requirement. CMOS has a +3 to +20V requirement; therefore, it is to be expected that a logic 1 in CMOS is supply voltage dependent. Actually, a CMOS logic 1 is approximately 0.7 times the supply voltage. The following formula defines the period of the oscillation, or how long it takes from rising edge to the successive rising edge for all three inverters. Frequency is the reciprocal of the period.

period = T
R = resistance expressed in ohms
C = capacitance expressed in Farads
$V_{TR}$ = threshold voltage of a logic 1, or 0.7 times $V_{DD}$ (the supply voltage)
$V_{DD}$ = supply voltage
$V_D$ = one diode voltage drop, or 0.7 volts

Assuming a 15 volt supply voltage,

$$T = -RC \ln \frac{(V_{TR})\, V_{DD} - V_{TR})}{(V_{DD} + V_D)^2}$$

$$T = -RC \ln \frac{(10.5)(15 - 10.5)}{(15 + 0.7)^2}$$

$$T = -RC \ln 0.19169$$

$$T = (-RC)\,(-1.65)$$

$$T = 1.65\ RC \qquad f = \frac{0.605}{RC}$$

**Q:** What is the formula for the frequency of oscillation with an astable inverter in the CMOS family?

**A:** $f = \dfrac{0.605}{RC}$

The three "inverter" astable multivibrators are shown as follows:

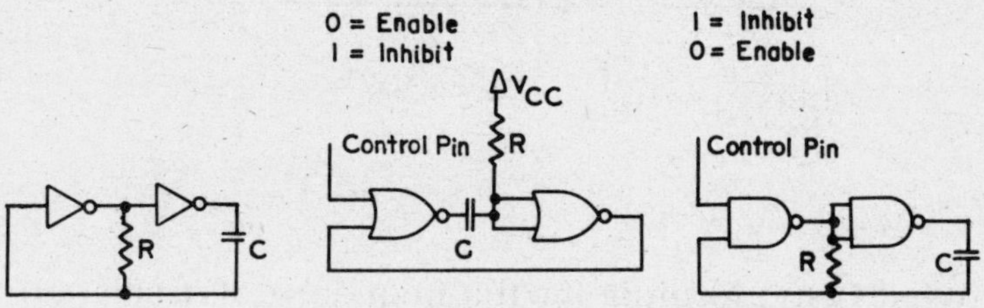


Note, however, that the NAND and NOR oscillators can be gated or turned on and off by controlling (enabling or inhibiting) the indicated pins.

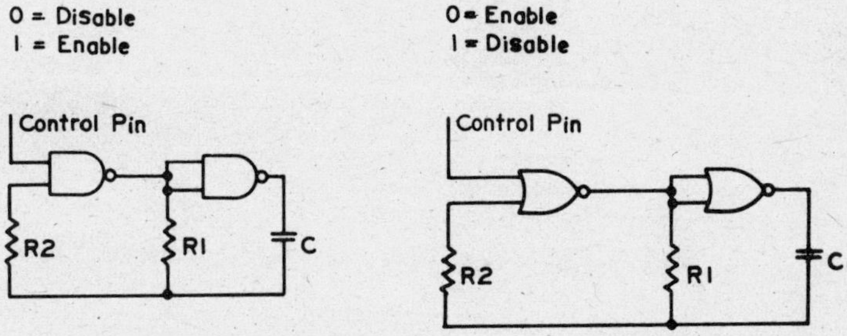


Refer back to the formula for the period "T" and note how the term $V_{DD}$, the supply voltage, is critical to the period and, therefore, frequency.
The presence of R2 compensates for this by making the RC combination more important in the determination of frequency.

**Q:** What is the purpose of R2?

**A:** To stabilize the frequency when the supply voltage changes.

The Exclusive-OR oscillator shown below has its frequency determined by the following formula:

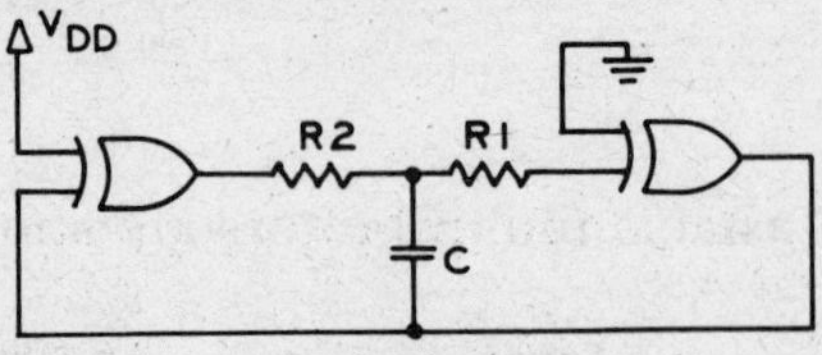


$$f = \frac{0.56}{RC} \qquad \text{when } R_1 = R_2$$

**Q:** What is the formula for the frequency of oscillation with an Exclusive-OR gate oscillator?

**A:** $f = \frac{0.56}{RC}$

**CHAPTER**

**4**

# Binary Principles

**OBJECTIVES:** Upon completion of Chapter 4 you will be able to do the following:

- ☐ List the two conditions a proposition can take
- ☐ Identify why the letters of the alphabet are used in digital electronics
- ☐ List the alternative terms for the opposite of a proposition
- ☐ Explain what combinations of propositions are called
- ☐ Explain how propositions are joined together
- ☐ Explain what the AND connective is called
- ☐ Explain what the OR connective is called
- ☐ Explain what two things determine the truth value of a sentence, expression, or a function
- ☐ Explain what makes a conjunction of two propositions true
- ☐ Write the inverse of the complex expression Y
- ☐ Construct truth tables for various gate circuits
- ☐ Develop Boolean expressions from truth tables
- ☐ Write Boolean expressions for various gate circuits

IN DIGITAL ELECTRONICS, three main theoretical ideas are put into practice: Boolean algebra and symbolic logic; binary counting; and codes. In electronics, the term "logic" refers to a limited application of symbolic logic and Boolean algebra. Both are very useful in decision-making situations; they also provide a very convenient method of expressing many physical problems found in industry, computers, and communications.

*Binary counting* is an alternative to our normal counting in the decimal system. It is much easier to implement electronically, and it is much faster and more reliable in operation. Information can be very conveniently coded in digital form. It can then be transmitted and manipulated by simple digital circuits.

What is fortunate is that these three areas can be implemented electronically with the same simple circuits. This chapter deals with each area separately.

# Logic

This section reviews the basics of Boolean algebra and symbolic logic which are used in digital electronics. The theorems are simply stated to show which are being used. None are proven as they would be in a book on mathematics. For a more complete coverage the student should read Ruth Ashley's *Background Math for a Computer World* in this series, and *Introduction to Finite Mathematics* by Kemeney, Thompson, and Snell, published by Prentice-Hall.

In this section we will review:

1. The basic propositions
2. The negation of propositions
3. Simple functions, expressions, or sentences using the Conjunction and the Disjunction, called here AND and OR
4. Truth tables
5. The symbols used
6. 1 and 0 as logic symbols
7. The negation of functions NAND and NOR
8. The truth table for these
9. How to invert functions with a truth table
10. How to turn expressions into a truth table
11. How to turn a truth table into an expression

Algebraic simplification of expressions and De Morgan's Laws are treated at the end of the chapter.

Symbolic logic is an algebra of propositions. Propositions are simple statements which are either true or false.

**Q:** What two conditions can a proposition take?

**A:** True or False.

There are no intermediate situations or conditions. Hence, symbolic logic is concerned with "two-valued" statements or propositions.

**Q:** What values other than true or false can a proposition take?

**A:** None. True and False are the only values.

**Q:** Can a proposition be both simultaneously?

**A:** No. It must be one or the other.

Once a proposition has been stated, it is common to represent it by a letter of the alphabet. This is a convenient shorthand. Thus, the proposition "The switch is closed" can be represented by the letter A.

**Q:** Why do we use letters of the alphabet to represent propositions?

**A:** For convenience.

The opposite of a statement or a proposition is often needed or used. This is called the *negation* of the proposition. The terms *not* and *inverse* are also used.

**Q:** What are the alternative terms for the opposite of a proposition?

**A:** NOT, negation, and inverse.

If a proposition is represented by the letter A, then its negation is given the letter A with a bar over it ($\overline{A}$).

**Q:** What is the inverse of A?

**A:** $\overline{A}$. This is called NOT A.

**Q:** If $\overline{A}$ was used to represent a closed switch, what would be used for an open switch?

**A:** A.

In symbolic logic, propositions are combined into complex forms called sentences, functions, or expressions.

**Q:** Propositions are often combined together. What are the combinations called?

**A:** Sentences, functions, or expressions.

Propositions are combined by the use of "connectives". Two or more propositions combined together by a connective form a sentence or a function or an expression.

**Q:** How are propositions joined together?

**A:** With connectives.

There are two connectives which are used extensively in electronics. The first is called the *conjunction* and it is indicated by the use of the word AND.

**Q:** What is the AND connective called?

**A:** The conjunction.

The other connective is called the *disjunction* and is indicated by the word OR.

**Q:** What is the OR connective called?

**A:** The disjunction.

When sentences are formed by combining propositions, the sentence has a "truth value". The truth value of the sentence depends upon two things:

1. The truth of the propositions in the sentence
2. The connectives used to make up the sentence

**Q:** What two things determine the truth value of a sentence, expression, or a function?

**A:** The truth of the propositions, and the connective used. A sentence formed from the conjunction of propositions is true *only if all the propositions are true.*

**Q:** What makes a conjunction of two propositions true?

**A:** The fact that both of the propositions used are true.

As an example, consider this statement: "Dogs have hair and birds have feathers."

**Q:** What two propositions have been used in this sentence?

**A:** Dogs have hair. Birds have feathers.

**Q:** What connective has been used?

**A:** The conjunction AND.

**Q:** Are both propositions true statements?

**A:** Yes.

**Q:** Is the sentence true?

**A:** Yes.

As another example, consider this statement: "Dogs have hair and birds have scales."

**Q:** Is this sentence true?

**A:** No.

**Q:** Why is it not true?

**A:** Because one of its propositions is not true.

A truth table is useful for listing the combinations of true and false that can be used to make up a function. The combinations shown here are for the AND.

| Proposition A | Proposition B | Function Value |
|---|---|---|
| F | F | |
| F | T | |
| T | F | |
| T | T | |

**Q:** Fill in the last column for the function value.

**A:** It is this:

F
F
F
T

A sentence formed from the disjunction of propositions is true if *any one or all of its propositions are true*.

**Q:** What makes the disjunction of two propositions true?

**A:** Any one, or all, of its propositions must be true.

As an example, consider this statement: "Dogs have hair or birds have scales."

**Q:** What two propositions have been used?

**A:** Dogs have hair. Birds have scales.

**Q:** What connective has been used?

**A:** The disjunction OR.

**Q:** Is one of the propositions a true statement?

**A:** Yes.

**Q:** Is the sentence a true statement?

**A:** Yes.

**Q:** Why?

**A:** Because one or more of the propositions are true.

A truth table is useful for listing the combinations of true and false that can be used to make up this function. The combinations shown on the next page are for the OR.

| Proposition A | Proposition B | Function Values |
|---|---|---|
| F | F | |
| F | T | |
| T | F | |
| T | T | |

**Q:** Fill in the last column for the function value.

**A:** This is it:

F
T
T
T

The disjunction is false only when both or all of its propositions are false. For example, "Dogs have wings or birds have hair."

**Q:** Is this sentence true?

**A:** No.

In arithmetic and ordinary algebra, symbols are used as instructions. For example, + means add, − means subtract, × means multiply. In Boolean algebra, symbols are also used to represent the conjunction and the disjunction. For the conjunction AND, the symbols × or • are used between the propositions. Often the propositions will be written close together with no symbol or space between them.

**Q:** Write the conjunction of A and B using the above.

**A:** A × B
A • B
AB

Each of these would be read as "A AND B".

The disjunction OR is indicated by the + symbol. No other symbol is used.

**Q:** How is the disjunction of A and B written?

**A:** A + B. This is read "A OR B".

Note the distinction made between the conjunction AND and the English language connective "and". The conjunction is always written in capitals, while the language connective is always written in lower case.

In ordinary algebra and arithmetic, the signs + and × mean addition and multiplication, respectively. That is not their meaning here.

Parentheses are also used in a manner similar to normal algebra. They help to organize the propositions into convenient logical groups. For example: A × (A+B). When propositions are combined to form sentences or expressions, a letter is also used to represent the expression formed. For example: X + A + B.

The = sign (equal sign) is used to separate the letter representing the expression from the proposition and connective symbols. The = sign is a convenience symbol. It does not necessarily mean "equal to". Its meaning changes with the problem or the statement. Usually it is clear from the conditions. The following examples will clarify this.

These propositions and statements are typical of those found in simple electronics.

**Example 1.** The lamp will light up when switch A or switch B is closed.

| | | | | |
|---|---|---|---|---|
| L | = | A | + | B |
| Lamp | will light | Switch A is closed | or | Switch B is closed |

**Example 2.** The gate will open when the handle at the top and the handle at the bottom are both pressed down.

| | | | | |
|---|---|---|---|---|
| G | = | T | × | B |
| Gate | will open | Top handle pressed | and | Bottom handle pressed |

**Example 3.** The tank will fill only when the top valve is open and the bottom valve is closed.

| $T$ | $=$ | $\overline{P}$ | $\times$ | $Q$ |
|---|---|---|---|---|
| Tank | will fill | Top valve is open | and | Bottom valve is closed |

(P is a closed switch so $\overline{P}$ is an open switch.)

At this point, it is convenient to introduce a simple idea which is used extensively in electronics. When truth tables are used, instead of T and F for True and False, the digits 1 and 0 are used. 1 is used to mean True. 0 is used to mean False.

**Q:** Re-do the truth tables for the AND using 1s and 0s.

**A:**

| Proposition A | Proposition B | Function Values |
|---|---|---|
| 0 | 0 | 0 |
| 0 | 1 | 0 |
| 1 | 0 | 0 |
| 1 | 1 | 1 |

**Q:** Repeat the truth table for the OR using 1s and 0s.

**A:**

| Proposition A | Proposition B | Function Values |
|---|---|---|
| 0 | 0 | 0 |
| 0 | 1 | 1 |
| 1 | 0 | 1 |
| 1 | 1 | 1 |

Later it will be shown how the 1 and the 0 are also used to indicate other conditions occurring in digital circuits.

It was stated earlier that every proposition has an inverse. This is also true of the complex expressions. They also have an inverse.

**Q:** If a complex expression is indicated by Y, what will its inverse be given by?

**A:** $\overline{Y}$.

Consider the simple expression involving the conjunction of two propositions:

$$Y = A \times B$$

**Q:** What is the inverse of this? Write it out.

**A:** $\overline{Y} = \overline{A \times B}$. Note that the bar is placed over the entire right-hand section. Note that it does not extend over the = sign.

The truth table for the inverse of a function can be easily obtained from the truth table of the function.

For instance, consider the truth table for the AND of two propositions.

| A | B | Y |
|---|---|---|
| 0 | 0 | 0 |
| 0 | 1 | 0 |
| 1 | 0 | 0 |
| 1 | 1 | 1 |

In order to make a truth table for the inverse of a function, another column is added. All the 1s in the Y column are now entered as 0s. All the 0s are entered as 1s. This is shown here.

| A | B | Y | $\overline{Y}$ |
|---|---|---|---|
| 0 | 0 | 0 | 1 |
| 0 | 1 | 0 | 1 |
| 1 | 0 | 0 | 1 |
| 1 | 1 | 1 | 0 |

This last column is the truth table for the inverted AND.

The inverse of the AND is called the "NOT AND". It is usually shortened to "NAND". This is very important in digital electronics.

Consider the simple expression for the disjunction of two propositions:

$$Y = A + B$$

**Q:** What is the inverse for this?

**A:** $\overline{Y} = \overline{A + B}$.

The truth table for the inverse can be found in the same way as for the AND function.

**Q:** What is the truth table for the OR of two propositions? Write it out.

**A:**

| A | B | Y |
|---|---|---|
| 0 | 0 | 0 |
| 0 | 1 | 1 |
| 1 | 0 | 1 |
| 1 | 1 | 1 |

**Q:** Now add in the extra column and write the inverse.

**A:**

| A | B | Y | $\overline{Y}$ |
|---|---|---|---|
| 0 | 0 | 0 | 1 |
| 0 | 1 | 1 | 0 |
| 1 | 0 | 1 | 0 |
| 1 | 1 | 1 | 0 |

The inverse of the OR is called the "NOT OR". It is usually shortened to "NOR". This is also very important in digital electronics.

The reasons for their importance will become apparent later. At this time, the student should become as familiar with these expressions as possible.

Up until now, this introduction and review has used only two propositions, A and B, in making up sentences. Obviously, there is no limit to how many can be used. No matter how many are used, the same rules continue to apply. Consider this example:

$$Y = ABC$$

**Q:** Write out the truth table for this. Hint: Remember to use binary counting.

**A:**

| A | B | C | Y |
|---|---|---|---|
| 0 | 0 | 0 | 0 |
| 0 | 0 | 1 | 0 |
| 0 | 1 | 0 | 0 |
| 0 | 1 | 1 | 0 |
| 1 | 0 | 0 | 0 |
| 1 | 0 | 1 | 0 |
| 1 | 1 | 0 | 0 |
| 1 | 1 | 1 | 1 |

Consider this example:

$$Y = A + B + C$$

**Q:** Write out the truth table for this.

**A:**

| A | B | C | Y |
|---|---|---|---|
| 0 | 0 | 0 | 0 |
| 0 | 0 | 1 | 1 |
| 0 | 1 | 0 | 1 |
| 0 | 1 | 1 | 1 |
| 1 | 0 | 0 | 1 |
| 1 | 0 | 1 | 1 |
| 1 | 1 | 0 | 1 |
| 1 | 1 | 1 | 1 |

Each of these expressions has an inverse.

**Q:** Write out the inverse for the AND expression, and write out the inverse column ($\overline{Y}$) in the truth table.

**A:** $\overline{Y} = \overline{A\ B\ C}$

| $\overline{Y}$ |
|---|
| 0 |
| 1 |
| 1 |
| 1 |
| 1 |
| 1 |
| 1 |
| 1 |

## How to Write the Truth Table for any Boolean Expression

Any Boolean expression can be put into truth table form.

The simple conjunctions of AND and OR have already been shown. The following procedure will allow the truth table of any complex expression to be written.

1. Draw columns for each separate proposition in the expression.
2. Fill in the first row with 0s.
3. Fill in the remaining rows by counting in binary until all 1s appear.
4. Look at each separate ANDed section of the expression. Consider all TRUE propositions to be 1s and all FALSE propositions to be 0s.
5. Find the horizontal rows which correspond to these combinations of 1s and 0s.
6. Place a 1 in the final column of the table in the rows which correspond to these rows.
7. Place 0s in the remaining places in the final column.

A simple example will illustrate this. Write out the truth table for:

$$Y = A\bar{B} + AB$$

1. Draw columns for each proposition.

| A | B | Y |
|---|---|---|

2. Fill in the first row with 0s.

| A | B | Y |
|---|---|---|
| 0 | 0 | |

3. Fill in the remaining rows by counting in binary until all 1s appear, or by beginning with a 0 at the top and then a 1 and keep alternating.

| A | B | Y |
|---|---|---|
| 0 | 0 | |
| 0 | 1 | |
| 1 | 0 | |
| 1 | 1 | |

4. Look at each separate ANDed section.

$A\overline{B}$ this is 10
$AB$ this is 11

5. Find the horizontal rows which correspond to each part of the expression $Y = A\overline{B} + AB$.

| A | B | Y | | |
|---|---|---|---|---|
| 0 | 0 | | this is $\overline{AB}$ | |
| 0 | 1 | | this is $\overline{A}B$ | |
| 1 | 0 | | this is $A\overline{B}$ | we need this |
| 1 | 1 | | this is $AB$ | we need this |

6. Place a 1 in the final column corresponding to these rows.

| A | B | Y |
|---|---|---|
| 0 | 0 | |
| 0 | 1 | |
| 1 | 0 | 1 |
| 1 | 1 | 1 |

7. Place 0s in the remaining places in the final column.

| A | B | Y |
|---|---|---|
| 0 | 0 | 0 |
| 0 | 1 | 0 |
| 1 | 0 | 1 |
| 1 | 1 | 1 |

This is the truth table for the expression:
$Y = A\overline{B} + AB$

This example follows this same procedure, but is written more compactly. Find the truth table for:
$Y = \overline{AB} + A\overline{B}$

| | A | B | Y |
|---|---|---|---|
| 2. | 0 | 0 | |
| 3. | 0 | 1 | |
| | 1 | 0 | |
| | 1 | 1 | |

4. $\overline{A}\overline{B}$ is 00
   $A\overline{B}$ is 10

| 5. | A | B | Y | |
|---|---|---|---|---|
| 6. | 0 | 0 | 1 | this is $\overline{A}\overline{B}$ |
| | 0 | 1 | | |
| | 1 | 0 | 1 | this is $A\overline{B}$ |
| | 1 | 1 | | |

| 7. | A | B | Y |
|---|---|---|---|
| | 0 | 0 | 1 |
| | 0 | 1 | 0 |
| | 1 | 0 | 1 |
| | 1 | 1 | 0 |

This is the truth table for:

$$Y = \overline{A}\overline{B} + A\overline{B}$$

In the next example, three propositions are used to form the function. The process is the same, but now one extra column is needed. In counting from all 0s in binary to all 1s, we find that eight rows are needed.

$$Y = \overline{A}\overline{B}C + \overline{A}BC + A\overline{B}\overline{C} + AB\overline{C}$$

| 1. | A | B | C | Y | |
|---|---|---|---|---|---|
| 2. | 0 | 0 | 0 | | |
| 3. | 0 | 0 | 1 | | this is $\overline{A}\overline{B}C$ |
| | 0 | 1 | 0 | | |
| | 0 | 1 | 1 | | this is $\overline{A}BC$ |
| | 1 | 0 | 0 | | this is $A\overline{B}\overline{C}$ |
| | 1 | 0 | 1 | | |
| | 1 | 1 | 0 | | this is $AB\overline{C}$ |
| | 1 | 1 | 1 | | |

4-7.

| A | B | C | Y |
|---|---|---|---|
| 0 | 0 | 0 | 0 |
| 0 | 0 | 1 | 1 |
| 0 | 1 | 0 | 0 |
| 0 | 1 | 1 | 1 |
| 1 | 0 | 0 | 1 |
| 1 | 0 | 1 | 0 |
| 1 | 1 | 0 | 1 |
| 1 | 1 | 1 | 0 |

The only problem likely to be encountered with this type of expression is when a proposition is missing from part of the expression. For example:

| Y = | AB | + | $\overline{BC}$ | + $\bar{A}BC$ |
|---|---|---|---|---|
| | C is missing | | A is missing | |

This is easily handled. AB has the proposition C missing. The C is replaced in both the True and False forms. So AB becomes ABC + AB$\bar{C}$. The $\overline{BC}$ is treated similarly. The A is replaced in both forms. So $\overline{BC}$ becomes A$\overline{BC}$ + $\overline{ABC}$.

Thus the total expression is:

$$Y = ABC + AB\bar{C} + A\overline{BC} + \overline{ABC} + \bar{A}BC$$

It is now ready to be put into truth table form. The procedure is exactly as before.

| A | B | C | Y | |
|---|---|---|---|---|
| 0 | 0 | 0 | 1 | this is $\overline{ABC}$ |
| 0 | 0 | 1 | 0 | |
| 0 | 1 | 0 | 0 | |
| 0 | 1 | 1 | 1 | this is $\bar{A}BC$ |
| 1 | 0 | 0 | 1 | this is $A\overline{BC}$ |
| 1 | 0 | 1 | 0 | |
| 1 | 1 | 0 | 1 | this is $AB\bar{C}$ |
| 1 | 1 | 1 | 1 | this is ABC |

This is the truth table for the expression:

$$Y = AB + \overline{BC} + \bar{A}BC$$

## How to Convert Truth Tables into Boolean Expressions

Every truth table is a representation of a Boolean expression, and the expression can be written down by reading the final column of a truth table. The final column is used in this manner:

- All the 1s in the final column represent ANDed propositions
- All the 0s in the final column are not used
- All the ANDed propositions will be ORed together

The procedure is the following:

1. Take the first row with a 1 in the final column.
2. Write the AND of the propositions which form this row.
3. Place a bar over those which have a 0 in their own column. Do not bar those which have a 1 in their own column.
4. Repeat this for all the other rows which have a 1 in the final column.
5. OR all of these together.

For example:

| A | B | Y |
|---|---|---|
| 0 | 0 | 0 |
| 0 | 1 | 1 |
| 1 | 0 | 1 |
| 1 | 1 | 1 |

1. Take the second row of the table.
2. Write the AND propositions which form this row: AB.
3. Place a bar over those which have a 0 in their own column: $\overline{A}B$.
4. 3rd row: $A\overline{B}$.
   4th row: AB.
5. OR all of these together: $Y = \overline{A}B + A\overline{B} + AB$.

Note that this procedure produces a correct Boolean expression but it is not the simplest expression. Usually it produces an expression which can be simplified considerably.

## How to Invert Boolean Expressions

Inverting a Boolean expression is very easy if the two previous techniques are combined. The procedure is very simple:

1. Write out the truth table of the expression.
2. Invert the last column by changing all the 1s to 0s and the 0s to 1s.
3. Write out the Boolean expression for this inverted last column.

This is the inverse of the original expression. Note that this does not produce the simplest or minimum inverse of the given expression. For this, simplification techniques must now be used.

Example:

Invert $Y = \overline{A}B + A\overline{B}$

1. Write out the truth table.

| A | B | Y | |
|---|---|---|---|
| 0 | 0 | 0 | |
| 0 | 1 | 1 | this is $\overline{A}B$ |
| 1 | 0 | 1 | this is $A\overline{B}$ |
| 1 | 1 | 0 | |

2. Invert the last column.

| A | B | Y | $\overline{Y}$ |
|---|---|---|---|
| 0 | 0 | 0 | 1 |
| 0 | 1 | 1 | 0 |
| 1 | 0 | 1 | 0 |
| 1 | 1 | 0 | 1 |

3. Write out the Boolean expression for this last column:

$$\overline{Y} = \overline{AB} + AB$$

# Binary Counting

Counting in decimal is the most familiar numbering system. We learn it from childhood. Implementing such counting by mechanical means is quite easy. The cash register and the mile counter in a car are two common examples. It is also easy to do electronically, and this has the advantage of being very much faster. In fact, the first electronic computers worked with the decimal number system.

However, the binary number system is much easier to use electronically, and in recent years its superiority has been established for use with computers, small calculators, and with automated equipment.

The system of binary counting is now familiar to most people as it has been taught in schools for several years. For the reader who is not familiar with it, Ruth Ashley's *Background Math for a Computer World* in this series is suggested.

In this section, we briefly review:

- ☐ Decimal numbers
- ☐ Binary numbers
- ☐ Binary counting
- ☐ Binary addition
- ☐ Binary subtraction
- ☐ Complements
- ☐ Decimal to Binary conversion
- ☐ Binary to Decimal conversion

In the decimal system of counting, ten symbols are used.

**Q:** What are they?

**A:** 0, 1, 2, 3, 4, 5, 6, 7, 8, 9.

These ten symbols are used to represent the number of objects we are counting. If the number is less than ten, then the appropriate symbol is used. When the number ten is reached, the same symbols are still used, but they are now written in two columns instead of one. For example: 10, 11, 12, etc.

In the decimal, the symbols in any column are worth ten times as much as the same symbol written in the column to the right. For example: 11, 22, 99, 111, 444, etc.

Using this idea, any number written with these symbols can be broken down as shown:

$$5555 = 5000 + 500 + 50 + 5$$

This idea can be extended to numbers which have different symbols in the columns:

$$4528 = 4000 + 500 + 20 + 8$$

Each column represents a power of ten which is multiplied by the number represented by the symbol:

$$4528 = 4 \times 1000 + 5 \times 100 + 2 \times 10 + 8$$
$$= 4 \times 10^3 + 5 \times 10^2 + 2 \times 10^1 + 8 \times 10^0$$

**Q:** Break these numbers down into columns and powers of ten: 1234, 5678, 9876, 1201, 2030.

**A:**
$$1234 = 1 \times 10^3 + 2 \times 10^2 + 3 \times 10^1 + 4 \times 10^0$$
$$5678 = 5 \times 10^3 + 6 \times 10^2 + 7 \times 10^1 + 8 \times 10^0$$
$$9876 = 9 \times 10^3 + 8 \times 10^2 + 7 \times 10^1 + 6 \times 10^0$$
$$1201 = 1 \times 10^3 + 2 \times 10^2 + 0 \times 10^1 + 1 \times 10^0$$
$$2030 = 2 \times 10^3 + 3 \times 10^1$$

The binary number system works in a similar manner. In this system, only two symbols are used: 0 and 1. These are used to represent the number of objects less than two. When two is reached, the same symbols are used but now they are written in columns, just as with the decimal system. For example: 10, 11, 110, 111, etc.

In this system, the symbols in any column are worth *twice* as much as the same symbol written in the column to the right. This is best illustrated by a table comparing the lower binary numbers with the lower decimal numbers.

| Decimal | Binary | Decimal | Binary |
|---|---|---|---|
| 0 | 0 | 9 | 1001 |
| 1 | 1 | 10 | 1010 |
| 2 | 10 | 11 | 1011 |
| 3 | 11 | 12 | 1100 |
| 4 | 100 | 13 | 1101 |
| 5 | 101 | 14 | 1110 |
| 6 | 110 | 15 | 1111 |
| 7 | 111 | 16 | 10000 |
| 8 | 1000 | | |

This can be continued indefinitely.

As with the decimal system, any number written in binary can be broken down as shown here:

$$11 = 10 + 1$$
$$10 = 10 + 0$$

The numbers in each column represent a power of two:

$$1101 = 1 \times 2^3 + 1 \times 2^2 + 0 \times 2^1 + 1 \times 2^0$$

**Q:** Break these numbers down into columns and powers of two:

1101   1011   1010   111   11

**A:**
$$1101 = 1 \times 2^3 + 1 \times 2^2 + 0 \times 2^1 + 1 \times 2^0$$
$$1011 = 1 \times 2^3 + 0 \times 2^2 + 1 \times 2^1 \times 1 \times 2^0$$
$$1010 = 1 \times 2^3 + 0 \times 2^2 + 1 \times 2^1 + 0 \times 2^0$$
$$111 = 1 \times 2^2 + 1 \times 2^1 + 1 \times 2^0$$
$$11 = 1 \times 2^1 + 1 \times 2^0$$

Binary numbers can be added. Binary addition is simple, and the basic rules are shown here. The decimal counterparts are shown below for comparison.

| | | | | | | |
|---|---|---|---|---|---|---|
| Binary: | 0 | 0 | 1 | 10 | 11 | 100 |
| | +0 | +1 | + 1 | + 1 | + 1 | + 1 |
| | 0 | 1 | 10 | 11 | 100 | 101 |
| Decimal: | 0 | 0 | 1 | 2 | 3 | 4 |
| | +0 | +1 | +1 | +1 | +1 | +1 |
| | 0 | 1 | 2 | 3 | 4 | 5 |

Note the carry into the next column.

**Q:** Add these binary numbers:

| | | | | | |
|---|---|---|---|---|---|
| 11 | 100 | 101 | 110 | 111 | 101 |
| 11 | 10 | 11 | 101 | 10 | 101 |

In all cases, check your answer with the familiar decimal numbers.

| | | | | | | |
|---|---|---|---|---|---|---|
| **A:** Binary: | 110 | 110 | 1000 | 1011 | 1001 | 1010 |
| Decimal: | 3 | 4 | 5 | 6 | 7 | 5 |
| | 3 | 2 | 3 | 5 | 2 | 5 |
| | 6 | 6 | 8 | 11 | 9 | 10 |

Binary numbers can also be subtracted. The simple subtraction rules are shown here. For comparison, the decimal equivalents are shown below.

| | | | | | | | | |
|---|---|---|---|---|---|---|---|---|
| Binary: | 0 | 1 | 1 | 10 | 11 | 11 | 100 | 100 |
| | −0 | −0 | −1 | − 1 | − 1 | −10 | − 10 | − 11 |
| | 0 | 1 | 0 | 1 | 10 | 1 | 10 | 1 |
| Decimal: | 0 | 1 | 1 | 2 | 3 | 3 | 4 | 4 |
| | −0 | −0 | −1 | −1 | −1 | −2 | −2 | −3 |
| | 0 | 1 | 0 | 1 | 2 | 1 | 2 | 1 |

As a safety precaution, subtraction should always be checked by readdition.

## Binary Complements

Binary subtraction is sometimes complicated, and is very difficult to achieve electronically; thus a method called *binary complements* is used. The 1s complement of a binary number is found by changing all the 1s to 0s, and all the 0s to 1s. This 1s complement of 101010 is 010101. Note the leading 0 in the complement is kept.

**Q:** Write the 1s complement of these:
11011 10101 10011 111101

**A:** 00100 01010 01100 000010

Complementing numbers electronically is very easy. This will be shown later. The 2s complement of a binary number is formed from the first complement. To form it, a 1 is added to the right-most digit, i.e., to the least significant digit.

Consider this number: 101011
The 1s complement is: 010100
The 2s complement is: 010101

**Q:** Find the 2s complement of these numbers:

| | | | | |
|---|---|---|---|---|
| 11011 | 10101 | 10011 | 111101 | 10100 |

**A:**

| | | | | |
|---|---|---|---|---|
| 00101 | 01011 | 01101 | 000011 | 01100 |

Subtraction can be performed very easily with complements. The 1s complement subtraction procedure is this:

1. Take the number to be subtracted (the subtrahend).
2. Form the 1s complement.
3. Add the 1s complement to the other number.
4. If the LH column produces a 1 as a carry, do two things:
   a. drop the 1.
   b. add a 1 to the right-most digit.

An example will clarify this. Subtract these two numbers: 111 − 101.

1. The subtrahend is 101.
2. The 1s complement is 010.
3. 
```
  111
 +010  (1s complement of 101)
 1001
```
4a. drop the carry-over (the left-most 1); this leaves
```
  001
    1    add 1 here; add these last two rows to get
  010    the answer
```
In decimal, this is 7 − 5 = 2.

**Q:** The following subtractions are given in decimal. Write them out in binary notation and perform the subtraction.

| | | | | | |
|---|---|---|---|---|---|
| 8 | 9 | 9 | 10 | 11 | 8 |
| −4 | −5 | −6 | −5 | −7 | −7 |
| 4 | 4 | 3 | 5 | 4 | 1 |

| | | | | | | |
|---|---|---|---|---|---|---|
| **A:** | 1000 | 1001 | 1001 | 1010 | 1011 | 1000 |
| | 0100 | 0101 | 0110 | 0101 | 0111 | 0111 |

Note how each of the numbers has been written with a leading 0. This is very important in this type of subtraction.

Now take the 1s complement of the lower number and add.

| | | | | | |
|---|---|---|---|---|---|
| 1000 | 1001 | 1001 | 1010 | 1011 | 1000 |
| +1011 | +1010 | +1001 | +1010 | +1000 | +1000 |
| 10011 | 10011 | 10010 | 10100 | 10011 | 10000 |

Drop the leading 1. Add 1 to the least significant digit.

| | | | | | |
|---|---|---|---|---|---|
| 0011 | 0011 | 0010 | 0100 | 0011 | 0000 |
| + 1 | + 1 | + 1 | + 1 | + 1 | + 1 |
| 100 | 100 | 11 | 101 | 100 | 1 |

The leading 0s can be dropped at this last line, which is the final answer. The answers should be checked with the decimal numbers above.

Now do the following exercises; the numbers are given in decimal. Convert to binary and perform the subtractions.

***Example 1***

| | | | | | |
|---|---|---|---|---|---|
| 4 | 5 | 7 | 9 | 12 | 15 |
| −3 | −1 | −6 | −4 | −8 | −13 |
| 1 | 4 | 1 | 5 | 4 | 2 |

Do not look until you have tried them yourself!

***Answer***

| | | | | | | |
|---|---|---|---|---|---|---|
| 100 | 101 | 111 | 1001 | 1100 | 1111 | top line |
| 100 | 110 | 001 | 1011 | 0111 | 0010 | 1s comps |
| 1000 | 1011 | 1000 | 10100 | 10011 | 10001 | |
| 000 | 011 | 000 | 0100 | 0011 | 0001 | leading 1 dropped |
| 1 | 1 | 1 | 1 | 1 | 1 | |
| 1 | 100 | 1 | 101 | 100 | 10 | answers |

The 2s complement can also be used for subtraction. The procedure is this:

1. Take the number to be subtracted.
2. Form the 2s complement.
3. Add it to the other number.
4. If a leading 1 is produced as a carry—ignore it.
5. The remaining number is the answer.
6. If a leading 1 is *not* produced as a carry, the number produced is the 2s complement of a negative answer. Then do the following:
   a. reduce to the 1s complement by subtracting a 1.
   b. turn the 1s complement into a binary number. This is the answer; do not forget the negative sign.

These following examples will clarify this.

***Example 2***

| Dec | Bin | 1sC | 2sC | |
|---|---|---|---|---|
| 7 | 111 | | 111 | |
| −5 | 101 | 010 | 011 | add |
| 2 | | | 1010 | |
| | | | | ignore the leading 1, leaving |
| | | | 010 | |
| | | | 010 | is a positive answer. |

***Example 3***

| | | | | |
|---|---|---|---|---|
| 13 | 1101 | | 1101 | |
| −10 | 1010 | 0101 | 0110 | add |
| 3 | | | 10011 | |
| | | | | ignore the leading 1, leaving |
| | | | 0011 | |
| | | | 11 | is a positive answer. |

***Example 4***

| Dec | Bin | 1sC | 2sC | |
|---|---|---|---|---|
| 10 | 1010 | | 1010 | |
| −13 | 1101 | 0010 | 0011 | add |
| 3 | | | 1101 | no leading 1, so this is a 2s complement form the 1s complement |
| | | | 1100 | |
| | | | | form the binary number |
| | | | 0011 | |
| | | | −11 | is the (negative) answer. |

***Example 5***

| | | | | |
|---|---|---|---|---|
| 5 | 101 | 101 | 101 | |
| −7 | 111 | 000 | 001 | add |
| −2 | | | 110 | no leading 1, so this is a 2s complement form the 1s complement |
| | | | 101 | |
| | | | | form the binary number |
| | | | 010 | |
| | | | −10 | is the (negative) answer. |

The student should now do the following exercises.

**Q:** The numbers are given in decimal form; convert to binary and perform the subtractions.

| | | | | | |
|---|---|---|---|---|---|
| 4 | 5 | 7 | 9 | 12 | 15 |
| −3 | −1 | −6 | −4 | − 8 | −13 |
| 1 | 4 | 1 | 5 | 4 | 2 |

Do not look until you have tried them yourself!

| | | | | | | | |
|---|---|---|---|---|---|---|---|
| **A:** | 100 | 101 | 111 | 1001 | 1100 | 1111 | |
| | 101 | 111 | 010 | 1100 | 1000 | 0011 | 2s comp |
| | 1001 | 1100 | 1001 | 10101 | 10100 | 10010 | |
| | 1 | 100 | 1 | 101 | 100 | 10 | |

**Q:** The student should now do these in the same way.

| | | | | | |
|---|---|---|---|---|---|
| 3 | 1 | 6 | 4 | 8 | 13 |
| −5 | −5 | −7 | −9 | −12 | −15 |
| −2 | −4 | −1 | −5 | −4 | −2 |

Again, do not look before you have solved them yourself.

**A:**

| | | | | | | |
|---|---|---|---|---|---|---|
| 011 | 001 | 110 | 0100 | 1000 | 1101 | top row |
| 011 | 011 | 001 | 0111 | 0100 | 0001 | 2s comp |
| 110 | 100 | 111 | 1011 | 1100 | 1110 | |
| 101 | 011 | 110 | 1010 | 1011 | 1101 | 1s comp |
| 010 | 100 | 001 | 0101 | 0100 | 0010 | answer |
| −2 | −4 | −1 | −5 | −4 | −2 | |

Note that all the columns are filled, even with a 0 in the leading column.

## Decimal to Binary Conversion

It is often necessary to convert numbers from one system to the other. Binary notation is needed for electronic work, but the decimal system is easier and more familiar to us. Two methods are shown.

### *Method 1*

The procedure is:

1. Write down the decimal number.
2. Write it as a sum of the largest power of two plus a remainder.
3. Write the remainder as a power of two plus a smaller remainder.
4. Continue in this until the remainder is 1 or 0.
5. Write out columns for each power of two.
6. Put a 1 in each column where a power of two was used in the above process. Put a 0 elsewhere.
7. The 1s and 0s will be the binary number.

***Example 1***

| | | |
|---|---|---|
| 19 = | 16 + 3 | 16 is largest power of two which is less than 19 |
| 3 = | 2 + 1 | 2 is largest power of 2 less than 3 |
| | | Stop as a 1 has appeared as a remainder. |

Write out columns of powers of two.

| | $2^5$ | $2^4$ | $2^3$ | $2^2$ | $2^1$ | $2^0$ |
|---|---|---|---|---|---|---|
| In decimal | 0 | 1 | 0 | 0 | 1 | 1 |
| these are: | | 16 | | | 2 | 1 |

The answer is binary 10011.

***Example 2***

63 = 32 + 31
31 = 16 + 15
15 = 8 + 7
7 = 4 + 3
3 = 2 + 1

| | $2^5$ | $2^4$ | $2^3$ | $2^2$ | $2^1$ | $2^0$ |
|---|---|---|---|---|---|---|
| In decimal | 1 | 1 | 1 | 1 | 1 | 1 |
| these are: | 32 | 16 | 8 | 4 | 2 | 1 |

So 63 in binary is 111111.

***Example 3***

This is an easier example, although the number is larger.

432 = 256 + 176
176 = 128 + 48
48 = 32 + 16
16 = 16 + 0

| | $2^8$ | $2^7$ | $2^6$ | $2^5$ | $2^4$ | $2^3$ | $2^2$ | $2^1$ | $2^0$ |
|---|---|---|---|---|---|---|---|---|---|
| In decimal | 1 | 1 | 0 | 1 | 1 | 0 | 0 | 0 | 0 |
| these are: | 256 | 128 | | 32 | 16 | | | | |

So 432 in binary is 110110000.

This process is very easy with practice, but it does rely on knowing the powers of 2.

**Q:** Convert the following decimal numbers to binary notation:
21, 43, 56, 78, 133, 257, 329.

**A:** 10101, 101001, 111000, 1001110, 10000101, 100000001, 101001001.

***Method 2***

This method is a little more formalized and does not require the student to know the powers of 2. It is a successive division process. The procedure is:

1. Divide the number by 2.
2. Write the dividend under the number.
3. Write the remainder at the right.
4. Now divide the new number by 2.
5. Again write the remainder at the right.
6. Continue until a 1 or a 0 is produced.
7. The 1s and 0s written as remainders are the binary number.

***Example 1***

| | | | *Remainder* |
|---|---|---|---|
| Divide 2 into 19 | 2 | ⌊19 | |
| Dividend | | 9 | 1 |
| Divide 2 into 9 | 2 | ⌊9 | |
| Dividend | | 4 | 1 |
| Divide 2 into 4 | 2 | ⌊4 | |
| Dividend | | 2 | 0 |
| Divide 2 into 2 | 2 | ⌊2 | |
| Dividend | | 1 | 0 |
| Divide 2 into 1 | 2 | ⌊1 | |
| This produces no dividend and the remainder is 1 | | 0 | 1 |

The binary number is read from the column of remainders, starting at the *bottom* and going *up*.

The answer is 10011.

***Example 2***

In this example, the numbers are written closer together. The process should be clear as it is the same as the previous example. Convert 63 into a binary number.

| | | *Remainder* |
|---|---|---|
| 2 | $\lfloor 63$ | |
| | 31 | 1 |
| 2 | $\lfloor 31$ | |
| | 15 | 1 |
| 2 | $\lfloor 15$ | |
| | 7 | 1 |
| 2 | $\lfloor 7$ | |
| | 3 | 1 |
| 2 | $\lfloor 3$ | |
| | 1 | 1 |
| 2 | $\lfloor 1$ | |
| | 0 | 1 |

The answer is 111111.

To conserve time and space, the writing-out process can be abbreviated slightly, as shown:

| | | |
|---|---|---|
| 2 | 63 | |
| 2 | 31 | 1 |
| 2 | 15 | 1 |
| 2 | 7 | 1 |
| 2 | 3 | 1 |
| 2 | 1 | 1 |
| | – | 1 |

The answer is 111111.

**Q:** Convert 78 into binary.

**A:**

| | | |
|---|---|---|
| 2 | 78 | |
| 2 | 39 | 0 |
| 2 | 19 | 1 |
| 2 | 9 | 1 |
| 2 | 4 | 1 |
| 2 | 2 | 0 |
| 2 | 1 | 0 |
| | – | 1 |

Reading the remainder column from bottom to top, the answer is 1001110.

**Q:** Convert 56 into binary.

**A:**

| | | |
|---|---|---|
| 2 | 56 | |
| 2 | 28 | 0 |
| 2 | 14 | 0 |
| 2 | 7 | 0 |
| 2 | 3 | 1 |
| 2 | 1 | 1 |
| 2 | 1 | 1 |

The answer is 111000.

## Binary to Decimal Conversion

Although various methods have been devised for binary to decimal conversion, only one will be shown here. This is by far the easiest method.

The procedure is this:

1. Write out the binary number as powers of two.
2. Convert each power of two into its decimal equivalent.
3. Add these to give the decimal number.

***Example 1***

Convert binary 11100101 to decimal.

| | $2^7$ | | $2^6$ | | $2^5$ | | $2^4$ | | $2^3$ | | $2^2$ | | $2^1$ | | $2^0$ |
|---|---|---|---|---|---|---|---|---|---|---|---|---|---|---|---|
| | 1 | | 1 | | 1 | | 0 | | 0 | | 1 | | 0 | | 1 |
| = | 128 | + | 64 | + | 32 | + | 0 | + | 0 | + | 4 | + | 0 | + | 1 |
| = | 229 | | | | | | | | | | | | | | |

Another way of writing this would be:

$$\begin{aligned} 11100101 &= 1\times2^7 + 1\times2^6 + 1\times2^5 + 0\times2^4 + 0\times2^3 + 1\times2^2 + 0\times2^1 + 1\times2^0 \\ &= 128 + 64 + 32 + 0 + 0 + 4 + 0 + 1 \\ &= 229 \end{aligned}$$

***Example 2***

Convert binary 101011 to decimal.

$$\begin{aligned} 101011 &= 1\times2^5 + 0\times2^4 + 1\times2^3 + 0\times2^2 + 1\times2^1 + 1\times2^0 \\ &= 32 + 0 + 8 + 0 + 2 + 1 \\ &= 43 \end{aligned}$$

The student should convert the following binary numbers to decimal, and check the answers obtained with previous examples in the text: 111000, 1001110, 111111, 10011, 1001110, 100000001.

# Binary Codes

Binary codes are patterns of 1s and 0s which are used to represent other information. Many codes have been devised and used. The choice of which code to use depends upon the particular job to be undertaken.

The subject of coding and transmitting information is very complex both mathematically and electronically, and a full treatise would require several volumes. However, the basic principles can be stated very simply and clearly. The purpose of this introduction is to make the student aware that 1) information is converted into digital form; that 2) this is the modern method of handling information and data; and 3) that various codes can be used.

The actual techniques of how the information is encoded, manipulated, transmitted, and decoded will not be covered in this book.

This section will discuss:

1. What type of information can be coded
2. A simple code
3. How errors are detected
4. Why codes are used

The most common type of information used in our lives can be represented by letters, numbers, punctuation marks, math symbols, etc. Hence, this is what is most needed to be encoded in a binary code for sending data from place to place. Typical examples are bank balances, credit card data, letters, telegrams, industrial data, etc.

For example, a customer in New York presents a credit card at a store, but his home is in Los Angeles. By punching the appropriate information into a terminal, the veracity of the card and the holder can be immediately checked by a computer in Chicago. In an automated environment, the user's location can be considered "transparent." A traveller in Los Angeles can reserve a plane seat in London. The seat information is typed into a terminal and the reservation accepted by the airline computer in New York.

A very early example of a digital code is the well-known Morse code. It was not called a digital code until quite recently, and it differs from modern digital codes in some important respects. Morse code is generated by closing and opening a switch, which sends a voltage down a telegraph line or over a radio. At the receiving end, a similar key opens and closes in the same pattern as the sender's key.

In Morse code, it is not the actual pattern of 1s and 0s which make the code but the time duration of the 1s. Dots are 1s of short duration and dashes are 1s of long duration. The 0s are the spaces between the 1s, and serve only to separate the 1s from each other.

In modern digital code, the 1s and the 0s are of the same duration and both are used to represent the information.

**Q:** What is the main difference between Morse and a modern digital code?

**A:** In Morse code, the information is contained in the 1s, which are of long and short duration. They are called Dots and Dashes. In a modern code, both the 1s and 0s carry the information, and they are of the same time duration.

A very simple modern code is the binary number system.

**Q:** Write the numbers 1–15 in this code.

**A:**

| | | |
|---|---|---|
| 0001 | — | 1 |
| 0010 | — | 2 |
| 0011 | — | 3 |
| 0100 | — | 4 |
| 0101 | — | 5 |
| 0110 | — | 6 |
| 0111 | — | 7 |
| 1000 | — | 8 |
| 1001 | — | 9 |
| 1010 | — | 10 |
| 1011 | — | 11 |
| 1100 | — | 12 |
| 1101 | — | 13 |
| 1110 | — | 14 |
| 1111 | — | 15 |

It would seem possible that by representing the letters of the alphabet by the numbers 1–26, words could also be transmitted with this code.

**Q:** Can you think of a difficulty with this?

**A:** How could you tell the numbers from the letters!

One easy way would be to precede all the letters with a 1 and all the numbers with a 0. For example, what does 01110 represent? The first digit is a 0, so it is a number. 01110 is binary 14; so this is the code for the number 14.

**Q:** What is this in the same code: 11110?

**A:** Since the first digit is a 1, this is a letter. The remaining digits, 1110, are binary 14. So this could be the fourteenth letter of the alphabet; that is, N.

If this idea was used, then the numbers from 27 onwards could be preceded with a 1 and used for special symbols such as punctuation and mathematics, etc. This is a simple example of how the binary numbers could be used to code a large amount of information.

The binary numbers do not have to be used. Many other patterns can be used, and are used, to represent the information.

When a code is sent over a line to another place, it is essential that it arrives with as few errors as possible. Why do we not want errors? Because wrong information is useless, or even worse, it can cause unnecessary trouble.

Unfortunately, errors are introduced into all electrical equipment and electrical signals. We are all familiar with noise and interference on radio, TV, and the telephone. Distortion in an audio record player or tapedeck is also a form of error.

**Q:** Can you think of some causes of electrical interference?

**A:** Storms, small appliances such as drills, vacuum cleaners, etc. Vehicles with bad ignition wiring also cause interference.

Most of these above sources of trouble will affect the AC, or analog, signals much more readily than they will affect a digital signal. So a digital signal is less prone to interference than an AC signal.

**Q:** Why is a digital signal safer than an AC signal for transmitting code?

**A:** It is less affected by the common noise sources.

When a digital code has errors, induced by noise or other causes, the errors are usually in the form of missing or added "bits"; that is, too many or not enough 1s will arrive. If these errors were allowed to go through and compound, then obviously the information would be useless when received.

Consider a traveller trying to buy an air ticket in London, so that he could return home to Los Angeles. The traveller's credit card may be indicated as stolen, overdrawn, or out of date because of errors received and transmitted by the computer.

To prevent such disasters, all digital codes have error-checking information added to them. This then becomes part of the code along with the information. The most common error check is called the *parity bit*. This is an extra 1 and 0 added to the code.

**Q:** What is a parity bit?

**A:** It is an extra 1 or 0 added to a digital code.

**Q:** What is it used for?

**A:** Detecting errors.

The manner in which a parity bit works is quite easy. In each individual piece of information in the code, the number of 1s is counted. If it is even, then a 0 is added to the end; if it is odd, then a 1 is added to the end:

| | | | |
|---|---|---|---|
| 101010 | — | three 1s are used | So add a 1 to give 1010101 |
| 101011 | — | four 1s are used | So add a 0 to give 1010110 |

This is called an *even parity* system, or code, as the number of 1s transmitted is always even. If the receiver counts the 1s and finds an even number, it assumes the incoming data is correct. If it finds an odd number, it assumes that an error has been received.

**Q:** Add parity bits to these digital codes. Use even parity.

101011 001001 111101 111010 101001 001101

**A:** 1010110 0010010 1111011 1110100 1010011 0011011

**Q:** Are these received codes correct or incorrect?

| | | | | |
|---|---|---|---|---|
| 1010110 | 1100101 | 1010111 | 1110001 | 0011001 |

**A:**

| | | | | |
|---|---|---|---|---|
| even | even | odd | even | odd |
| correct | correct | wrong | correct | wrong |

Odd parity can also be used.

**Q:** What do you think odd parity is?

**A:** An odd number of 1s is transmitted.

The manner in which the sending equipment adds a parity bit will not be covered in this book; nor how the receiving equipment counts the bits and determines which bit is incorrect and corrects it. The answers lie in the use of statistics and information theory. This is far too complex to cover here, but a simple example will illustrate the principle on which it is based.

Assume a "Q" has been received and correctly identified. The next letter received has an error in its code. What is that letter most likely to be? In the English language a "U" is statistically most likely to follow a "Q". So this is automatically inserted into the information in the hope it is correct.

The most used letter in the English alphabet is "E". When in doubt, this is often inserted. Usually this results in misspelled telegrams but nothing more serious. For very important data where errors cannot be tolerated, such as the remote controls for space vehicles, more complex codes are used.

Two very important characteristics of digital codes are that they can be transmitted very fast, and they can contain large amounts of information. A 10-minute recorded telephone conversation can be encoded and transmitted in a few seconds.

**Q:** What are two important characteristics of digital codes?

**A:** Large information contents and high speed.

This high speed and large information content are not separate qualities. They are very interdependent. A study of this leads to some mathematical complexities which we will not cover; they are usually referred to as "Nyquist criteria", after the mathematician who first studied information theory.

Binary codes are used as they are a very convenient way to represent many different types of information. It is easy to transform most forms of data into a binary code; and once it is in that form, it is easy to protect, to check and eliminate errors, and to keep it secret if desired. Above all, binary codes are easy to implement, manipulate, and to store with high-speed digital electronic equipment.

**Q:** Give some reasons why binary codes are used.

**A:** Binary codes are easy to use and implement, easy to store data, to transmit data, and to check errors, etc.

Because of the ease of using codes, they have become very common. It has been said we are moving away from the established analog representations of information and toward a binary or digital information society. A simple example is the telephone.

Most conversations are not sent as an AC signal, which is an analog signal directly representing the voice, but are turned into a digital code. The code is converted back to an AC signal by the receiving exchange for use in the called person's telephone.

A more spectacular example is the TV pictures from Mars and Jupiter. The output from the camera was converted to a binary code for transmission back to Earth. On receipt, the errors were removed by computer and then the code converted into a picture.

It has been stated in all seriousness that if we ever contact extraterrestial life, it will probably be achieved by using a binary code. Many codes have been devised and used. The most common have been binary counting, the BCD code, the GRAY code, and the XS3 code. Two of the most important modern codes are the ASCII code, which is used in teletypes for most commercial applications, and the new "supermarket" code, which is a pattern of bars printed on jars and packets of consumer goods. This pattern is read optically or magnetically and is translated by a small computer into the price of the item. Simultaneously, it adjusts the inventory, computes the sales tax, decides if food stamps are acceptable, and keeps a running total of the day's transactions. The student should be aware that codes exist and that they are used extensively.

## Algebraic Simplification

The simplification of complex Boolean expressions is a common mathematical exercise which has an important counterpart in digital electronics. The simpler the expression or function, the simpler the circuit needed to represent it.

A common procedure in digital design is first to express a problem in Boolean or symbolic form, and then to simplify the expression. Once the expression is reduced to its minimum state, then the circuit is designed. Here we will review the most common formulas and theorems used in simplification.

The main identities used are:

| | |
|---|---|
| $A + 0 = A$ | $A \bullet 1 = A$ |
| $A + 1 = 1$ | $A \bullet 0 = 0$ |
| $A + A = 1$ | $A \bullet A = A$ |
| $A + \overline{A} = 1$ | $A \bullet \overline{A} = 0$ |
| $A + AB = A$ | $A(A + B) = A$ |
| $A\overline{B} + B = A + B$ | $(A + \overline{B})B = AB$ |
| $AB + A\overline{B} = A$ | $(A + B)(A + \overline{B}) = A$ |
| | $(A + B)(A + C) = A + BC$ |

$AB + BC + \overline{A}C = AB + \overline{A}C$

No attempt will be made to prove these identities in a strictly mathematical sense. However, a truth table does demonstrate the validity of the identities, and a technique will be shown for this.

The technique is to first build the truth table for the left-hand side of the identity. Then build the table for the right-hand side.

Compare the final columns of the tables. If they are the same, then obviously both sides of the expression are the same.

***Example 1***

Show $A\overline{B} + B = A + B$.

Table for LHS (left-hand side)

| A | B | $\overline{B}$ | $A\overline{B}$ | $A\overline{B}+B$ |
|---|---|---|---|---|
| 0 | 0 | 1 | 0 | 0 |
| 0 | 1 | 0 | 0 | 1 |
| 1 | 0 | 1 | 1 | 1 |
| 1 | 1 | 0 | 0 | 1 |

Table for RHS (right-hand side)

| A | B | A+B |
|---|---|---|
| 0 | 0 | 0 |
| 0 | 1 | 1 |
| 1 | 0 | 1 |
| 1 | 1 | 1 |

Compare the last two columns; they are the same. Hence, the expression $A\overline{B} + B = A + B$ is true.

In this next example, the same idea is used, only all the columns are written together.

***Example 2***

Show $(A + \overline{B})B = AB$.

| A | B | $A+\overline{B}$ | LHS $(A+\overline{B})B$ | RHS $AB$ |
|---|---|---|---|---|
| 0 | 0 | 1 | 0 | 0 |
| 0 | 1 | 0 | 0 | 0 |
| 1 | 0 | 1 | 0 | 0 |
| 1 | 1 | 1 | 1 | 1 |

This again shows they are the same.

These two theorems are of vital importance in digital electronics. They are known as De Morgan's Theorems.

Two of them are stated here:

1. $\overline{AB} = \overline{A} + \overline{B}$
2. $\overline{A + B} = \overline{A} \bullet \overline{B}$

These are the simplest forms of the two theorems, as they use two propositions only.

**Q:** Prove the first one with a truth table.

**A:**

| A | B | $\overline{A}$ | $\overline{B}$ | $AB$ | $\overline{AB}$ | $\overline{A} + \overline{B}$ |
|---|---|---|---|---|---|---|
| 0 | 0 | 1 | 1 | 0 | 1 | 1 |
| 0 | 1 | 1 | 0 | 0 | 1 | 1 |
| 1 | 0 | 0 | 1 | 0 | 1 | 1 |
| 1 | 1 | 0 | 0 | 1 | 0 | 0 |

The last two columns are the same.

**Q:** Prove the second with a truth table.

**A:**

| A | B | $\overline{A}$ | $\overline{B}$ | $A+B$ | $\overline{A+B}$ | $\overline{A}\bullet\overline{B}$ |
|---|---|---|---|---|---|---|
| 0 | 0 | 1 | 1 | 0 | 1 | 1 |
| 0 | 1 | 1 | 0 | 1 | 0 | 0 |
| 1 | 0 | 0 | 1 | 1 | 0 | 0 |
| 1 | 1 | 0 | 0 | 1 | 0 | 0 |

The last two columns are the same.

These two theorems can be extended to cover any number of propositions. Here they are shown for three propositions:

1. $\overline{ABC} = \overline{A} + \overline{B} + \overline{C}$
2. $\overline{A + B + C} = \overline{A} \bullet \overline{B} \bullet \overline{C}$

As an exercise, the reader should show these are true by using truth tables. Also as an exercise, the reader should examine these two expressions and see if they are true:

1. $\overline{A + B} = \overline{A} + \overline{B}$
2. $\overline{A \bullet B} = \overline{A} \bullet \overline{B}$

To see if 1. is true, draw up a truth table for both sides.

| A | B | A + B | $\overline{A + B}$ | $\overline{A}$ | $\overline{B}$ | $\overline{A} + \overline{B}$ |
|---|---|---|---|---|---|---|
| 0 | 0 | 0 | 1 | 1 | 1 | 1 |
| 0 | 1 | 1 | 0 | 1 | 0 | 1 |
| 1 | 0 | 1 | 0 | 0 | 1 | 1 |
| 1 | 1 | 1 | 0 | 0 | 0 | 0 |

**Q:** Are these two columns the same?

**A:** No. They are not. Hence $\overline{A + B} \neq \overline{A} + \overline{B}$.

To see if 2. is true, draw up a truth table again.

| A | B | AB | $\overline{AB}$ | $\overline{A}$ | $\overline{B}$ | $\overline{A} \bullet \overline{B}$ |
|---|---|---|---|---|---|---|
| 0 | 0 | 0 | 1 | 1 | 1 | 1 |
| 0 | 1 | 0 | 1 | 1 | 0 | 0 |
| 1 | 0 | 0 | 1 | 0 | 1 | 0 |
| 1 | 1 | 1 | 0 | 0 | 0 | 0 |

**Q:** Are these two columns the same?

**A:** No. They are not. Hence, $\overline{AB} \neq \overline{A} \bullet \overline{B}$.

These last two examples should be studied very carefully. They are the cause of many mistakes in the early stages of this subject.

In text books about Boolean algebra, two main types of examples are found:

1. Proving the equivalence of the left and right sides of an expression
2. Simplifying

An example of the first type is:

Prove $A + AB = A$

An example of the second type is:

Simplify $Y = \overline{A}BC + AB\overline{C} + ABC$

Both of these situations can be handled using:

1. The standard forms given at the beginning of this section
2. De Morgan's Laws
3. The truth table method of writing the expression out

The following exercises will illustrate these techniques.

**Q:** Simplify:

1. $Y = A + \overline{A}B$
2. $Y = (A\overline{B} + B)\,\overline{(A + B)}$
3. $Y = (A + \overline{B} + \overline{C})\,(A + \overline{B}C)$
4. $Y = A + \overline{C} + B\,\overline{(A + C)} + (A + \overline{C})\,(A + B + \overline{C})$
5. $Y = B\overline{(C + D)} + AB\,\overline{(C + D)}$
6. $Y = (A\overline{C}D + A\overline{BC})(A\overline{C}D + A\overline{BC}) + \overline{AB}$

**A:**

1. $A$
2. $0$
3. $A + \overline{B}C$
4. $A + B + \overline{C}$
5. $B\,\overline{(C + D)}$
6. $\overline{AB}$

### *Exercises*

1. Show all the other identities given are true. Use the truth table method.
2. Show A(B + C) = AB + AC
3. Show AB + BC + B$\overline{AC}$ = B
4. Show (A$\overline{B}$ + B)(AB + A$\overline{B}$) = A

Take care in drawing out the truth table columns, and these will present no difficulty.

All identities can be shown to be true with this method.

### *Answers*

2.

| A | B | C | B+C | AB | AC | AB+AC | A(B+C) |
|---|---|---|---|---|---|---|---|
| 0 | 0 | 0 | 0 | 0 | 0 | 0 | 0 |
| 0 | 0 | 1 | 1 | 0 | 0 | 0 | 0 |
| 0 | 1 | 0 | 1 | 0 | 0 | 0 | 0 |
| 0 | 1 | 1 | 1 | 0 | 0 | 0 | 0 |
| 1 | 0 | 0 | 0 | 0 | 0 | 0 | 0 |
| 1 | 0 | 1 | 1 | 0 | 1 | 1 | 1 |
| 1 | 1 | 0 | 1 | 1 | 0 | 1 | 1 |
| 1 | 1 | 1 | 1 | 1 | 1 | 1 | 1 |

3.

| A | B | C | AB | BC | AC | $\overline{AC}$ | B$\overline{AC}$ | AB+BC+B$\overline{AC}$ |
|---|---|---|---|---|---|---|---|---|
| 0 | 0 | 0 | 0 | 0 | 0 | 1 | 0 | 0 |
| 0 | 0 | 1 | 0 | 0 | 0 | 1 | 0 | 0 |
| 0 | 1 | 0 | 0 | 0 | 0 | 1 | 1 | 1 |
| 0 | 1 | 1 | 0 | 1 | 0 | 1 | 1 | 1 |
| 1 | 0 | 0 | 0 | 0 | 0 | 1 | 0 | 0 |
| 1 | 0 | 1 | 0 | 0 | 1 | 0 | 0 | 0 |
| 1 | 1 | 0 | 1 | 0 | 0 | 1 | 1 | 1 |
| 1 | 1 | 1 | 1 | 1 | 1 | 0 | 0 | 1 |

4.

| A | B | $\overline{B}$ | A$\overline{B}$ | A$\overline{B}$+B | AB | AB+A$\overline{B}$ | (A$\overline{B}$+B)(AB+A$\overline{B}$) |
|---|---|---|---|---|---|---|---|
| 0 | 0 | 1 | 0 | 0 | 0 | 0 | 0 |
| 0 | 1 | 0 | 0 | 1 | 0 | 0 | 0 |
| 1 | 0 | 1 | 1 | 1 | 0 | 1 | 1 |
| 1 | 1 | 0 | 0 | 1 | 1 | 1 | 1 |

CHAPTER

# 5

# Basic Logic Gates

**OBJECTIVES:** Upon completion of Chapter 5 you will be able to:

- ☐ Build simple logic circuits using mechanical switches
- ☐ Write the signal for "The switch is open"
- ☐ Draw an AND gate circuit using switches
- ☐ Draw an OR gate circuit using switches
- ☐ Write a Boolean expression for an AND gate and an OR gate
- ☐ Define an inverting buffer
- ☐ Define a noninverting buffer
- ☐ Describe how inversion is indicated on a drawing symbol
- ☐ Draw the circuit symbol for an OR gate, followed by the symbol for an inverter
- ☐ Work several experiments using inverters, AND, OR, NAND, and NOR gates

MANY SIMPLE DIGITAL CIRCUITS are physical examples of the use of Boolean algebra. The circuits can be built directly from Boolean equations, thus enabling theoretical ideas to be transformed into practical reality.

All digital circuitry is based upon the use of simple circuits which perform the AND, OR, Exclusive-OR and the inversion functions.

In this chapter the student will learn:

1. How mechanical switches are used to perform logic functions
2. How diodes can be used for logic functions
3. What an inverter is
4. How a NAND is achieved
5. How a NOR is achieved
6. How circuits which follow the rules of Boolean algebra are said to "perform logic"

Throughout this book you will encounter numerous transistor circuits and, in a few instances, diodes as part of these circuits. These circuits are really logic gates; and although in real life you will be predominately concerned with gates appearing within IC packages, these have been included for your edification. They also serve as reinforcement of certain digital principles. You do not need any extensive knowledge of transistors; in fact, you really do not even need to know how a transistor works, if you only read the following greatly simplified explanation.

A transistor is a three-lead device made of semiconductor material with a base, emitter, and collector. It is represented in schematic form as follows:

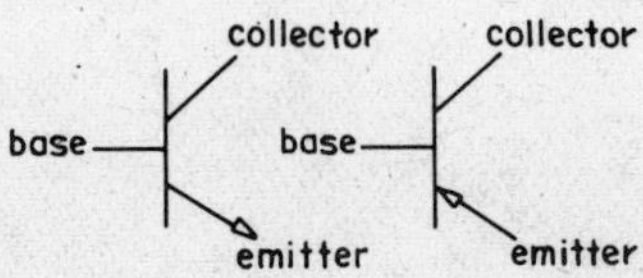


NPN Transistor PNP Transistor

The difference in bipolar transistors (the kind we are dealing with) is in the collector-to-emitter junction, or where the two coming together have different polarities. That is, on the NPN, the emitter points out. An easy way to remember is *N*ot *P*ointing i*N* (NPN)! Conversely, the PNP has an opposite polarity. Current flows from a more positive to a less positive source or potential. Therefore, in the NPN transistor current flows from the collector to the emitter. We will only be using NPN transistors in this book.

If the flow of current is equated to the flow of water, then we can further equate the base to the controlling device or to a faucet. As more current flows through the base, the transistor conducts more, or is turned on. In turn, this current flow through the collector resistor means that an IR drop (current × resistance of the collector resistor) occurs. At full conduction or turn on, approximately 98% of the total voltage is dropped across the collector resistor, leaving an almost zero voltage drop across the transistor. This constitutes a logic "0" or low because of the near-ground potential of voltage. Therefore, if the base has a high on it permitting base current flow, then the collector will have a low output.

Conversely, if the base has a logic low on it, either no current or an insufficiently small current will flow to cause the collector to allow current to flow, and this in turn will cause the majority of applied voltage to appear across the transistor.

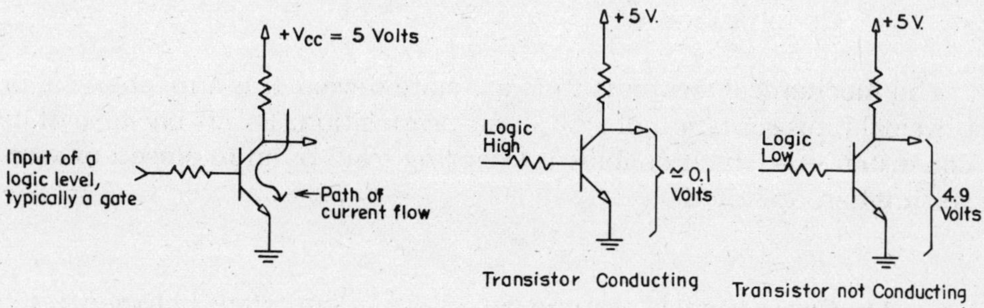


The diodes you will encounter are two-lead semiconductor devices whose purpose can best be described as current steerers or guiders. A diode has an anode and a cathode.

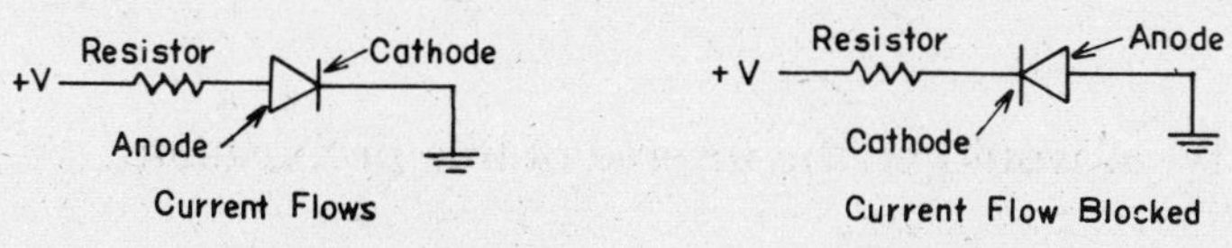


If the diode has a positive potential on its anode, the current in the diode flows in the direction of the arrow from anode to cathode. You may recall that a transistor's collector-to-emitter junction is just like this, and that current flows in the direction again of the arrow. If the anode is less positive than the cathode, then the diode acts as a blocker of current, in which case the full-applied voltage is applied or appears across it. Diodes are sometimes used in this passing/blocking manner to control what appears at the input of a transistor or on its base.

The simplest component which can be used in logic circuits is the mechanical switch.

**Q:** What component can be used to build a simple logic circuit?

**A:** The mechanical switch.

A mechanical switch has only two states in which it can be.

**Q:** What are they?

**A:** Open or closed.

The mechanical switch is thus a 2-state device. It is a mechanical or physical representation of a 2-valued proposition. This is because of its construction—which enables it to spring from open to closed or vice-versa when operated.

**Q:** Can a switch assume an intermediate state between open and closed?

**A:** No. It is either open or closed. There is no other condition.

The statement, "The switch is closed," is a simple example of a proposition. Naturally this proposition has an inverse.

**Q:** What would be the inverse of this proposition?

**A:** The switch is open.

A letter can be used to represent the proposition "The switch is closed." We will use the letter "A" for this. The inverse can also be represented by a letter. We will use the same letter with a bar over it.

**Q:** Write the symbol for "The switch is open."

**A:** $\overline{A}$.

**Q:** What is $\overline{A}$ called?

**A:** Not A.

The following figures illustrate these conditions.

**Q:** What does A mean?

**A:** The switch is closed.

**Q:** What does $\overline{A}$ mean?

**A:** The switch is open.

A truth table can be drawn which illustrates this.

| Condition of switch | Opposite condition of switch |
|---|---|
| Open | Closed |
| Closed | Open |

**Q:** Re-do this truth table using T for closed and F for open.

**A:**

| Condition | Opposite Condition |
|---|---|
| F | T |
| T | F |

Mechanical switches can be connected together in series and parallel connections. Consider the situation where a lamp is to be illuminated if two switches are closed. It is to remain off otherwise. Two switches connected in series will achieve this.

**Q:** Draw a simple circuit which will do this.

**A:** Your drawing should look something like this:

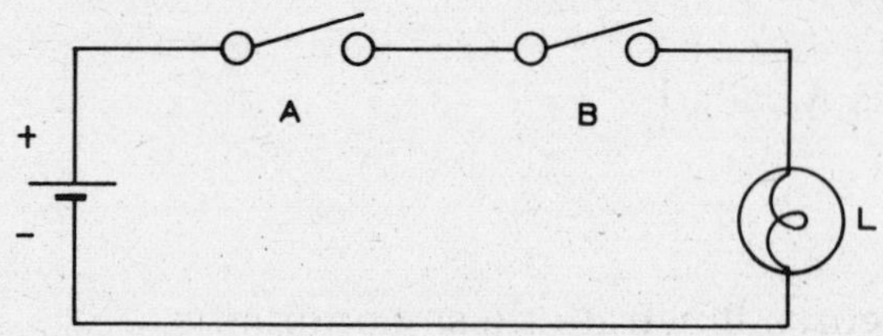


**Q:** In this above circuit, which switches must be closed to illuminate the lamp?

**A:** Both of them. Closing just one will not work.

A simple formal statement in Boolean algebra of the requirements stated above would be: The lamp is to be illuminated when both switch A is closed and switch B is closed.

**Q:** Can you write a Boolean expression for this? Explain what the letters and symbols mean.

**A:** L = AB
where L means the Lamp
= means the conditions under which the lamp is illuminated are given on the opposite side of this sign
A means switch A is closed
B means switch B is closed
AB means switches A AND B are both closed

The expression L = AB is a symbolic or Boolean statement, which means the lamp is on when both the switches are closed.

**Q:** What type of Boolean expression have we built with this simple circuit?

**A:** The AND expression.

In summary, you have seen how the AND can be represented by a simple series circuit of mechanical switches. A truth table can be used to describe this situation. Let T mean the switch is closed, and F mean it is open. The four conditions which can occur are set out in this table. "Sw" stands for "switch."

| Sw A | Sw B | Lamp |
|---|---|---|
| F | F | |
| F | T | |
| T | F | |
| T | T | |

**Q:** Fill in the last column for the condition of the lamp. Use F for a dark lamp, and T for when it is illuminated.

**A:**

| Sw A | Sw B | Lamp |
|---|---|---|
| F | F | F |
| F | T | F |
| T | F | F |
| T | T | T |

**Q:** Repeat this table, using 0 instead of F and 1 instead of T.

**A:**

| Sw A | Sw B | Lamp |
|---|---|---|
| 0 | 0 | 0 |
| 0 | 1 | 0 |
| 1 | 0 | 0 |
| 1 | 1 | 1 |

Two things should be observed in this table:

1. It is the truth table for the AND
2. The 0s and 1s are progressing in a binary count

Consider now the situation where a lamp is to be illuminated if any one or both of two switches are closed. It is to be off only when both switches are open. Two switches connected in parallel will achieve this.

**Q:** Draw a simple circuit which will do this.

**A:** Your drawing should be similar to this:

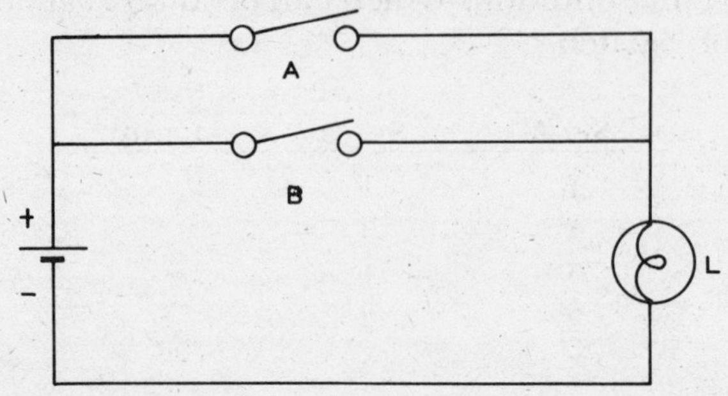


**Q:** In the above circuit, which switches must be closed to illuminate the lamp?

**A:** Any one of them by itself, or both of them together.

A simple formal statement of the requirements in Boolean algebra would be: The lamp is to be illuminated when switch A or switch B or when both are closed.

**Q:** Can you write a Boolean expression for this? Explain what the letters and symbols mean.

**A:** L = A + B
where L means the lamp
= means the condition under which the lamp is illuminated are given on the opposite side of this sign
A means switch A is closed
B means switch B is closed
A + B means switches A OR B are closed

The expression L = A + B is a symbolic statement or Boolean statement, which means the lamp is illuminated when both one switch or the other or both are closed.

**Q:** What type of Boolean expression have we built with this simple circuit?

**A:** The OR expression.

In summary: The OR can be represented by a simple parallel circuit of mechanical switches.

A truth table can be used to describe this situation. Let T mean the switch is closed and F mean it is open. The four conditions which can occur are set out in this table.

| Sw A | Sw B | Lamp |
|---|---|---|
| F | F | |
| F | T | |
| T | F | |
| T | T | |

**Q:** Fill in the last column for the condition of the lamp. Use F for a dark lamp and T for when it is illuminated.

**A:**

| Sw A | Sw B | Lamp |
|---|---|---|
| F | F | F |
| F | T | T |
| T | F | T |
| T | T | T |

**Q:** Repeat this table using 0 instead of F and 1 instead of T.

**A:**

| Sw A | Sw B | Lamp |
|---|---|---|
| 0 | 0 | 0 |
| 0 | 1 | 1 |
| 1 | 0 | 1 |
| 1 | 1 | 1 |

Two things should be observed in this table:

1. It is a truth table for the OR.
2. The 0s and 1s are progressing in a binary count.

The advantage of the truth tables is that they are not limited to switches and lamps, but can be used for all types of circuits and problems. It is a general process and is very useful in both expressing and designing circuits.

In the previous two examples, the 1 and the 0 were used to represent the conditions of the switches.

**Q:** What did the 1 mean?

**A:** A closed switch.

The 0 and the 1 can be used to represent other ideas. Consider the simple circuit shown here.

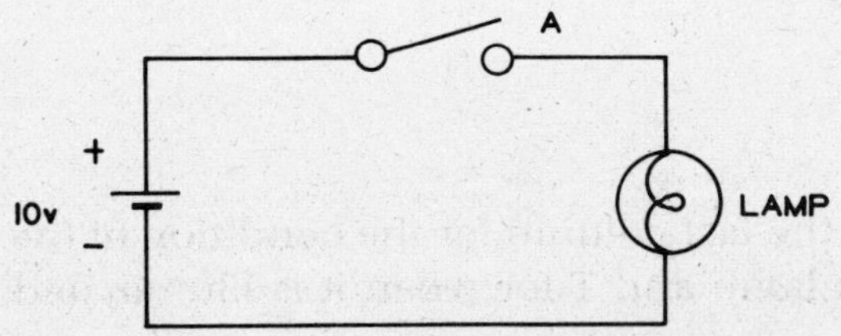


When the switch is closed, the battery voltage appears at the point marked A.

**Q:** What is the voltage level?

**A:** 10V.

When the switch is open, the voltage at this point is 0V. If a 1 is used to represent the closed switch, then a 1 will also indicate that 10V is present at point A.

**Q:** Why does a 1 also indicate 10V at point A?

**A:** Because a 1 means the switch is closed, thus connecting point A to the positive terminal of the battery.

Similarly the 0 will indicate that point A is at 0V. Instead of using the two digits 1 and 0 to represent the condition of the switch, we can also use them to indicate the presence or absence of a voltage at a given point.

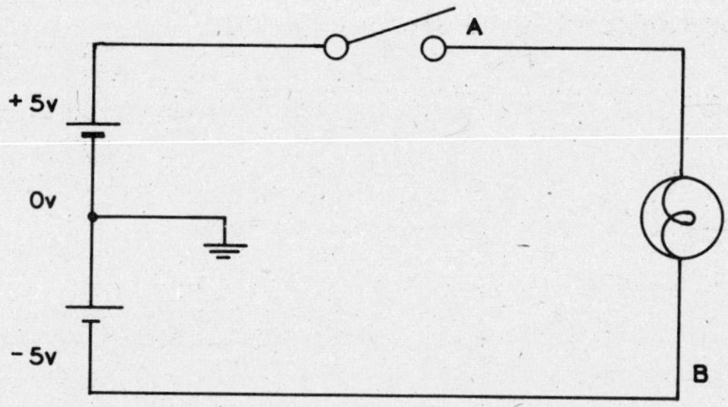


**Q:** When the switch is closed, what is the voltage at point A?

**A:** +5V.

**Q:** What is the voltage at point B?

**A:** It is always −5V.

Using the 1 and 0 here would give this:

1 means +5V
0 means −5V

Using the 1 and the 0 to represent the voltage levels is advantageous because it does not matter where the voltage comes from or how it is supplied. In the above examples, the voltages are supplied from a battery through a mechanical switch. There are many other ways of supplying voltage to a point in a circuit. Often, it is not necessary to know where the volts come from; just whether they are there or not.

The simple series and parallel circuits shown above can be combined into more complicated circuits. The following examples show simple combinations of switches which are used to illuminate a lamp.

**Q:** What two simple circuits are we discussing? What are we going to do with them?

**A:** Series and parallel circuits using mechanical switches. We are going to combine them in various ways to illuminate a lamp.

Look at this simple circuit. *NOTE*—Many times a circuit is analyzed and the complete circuit isn't shown. When this occurs in this chapter, assume that it is connected as shown below.

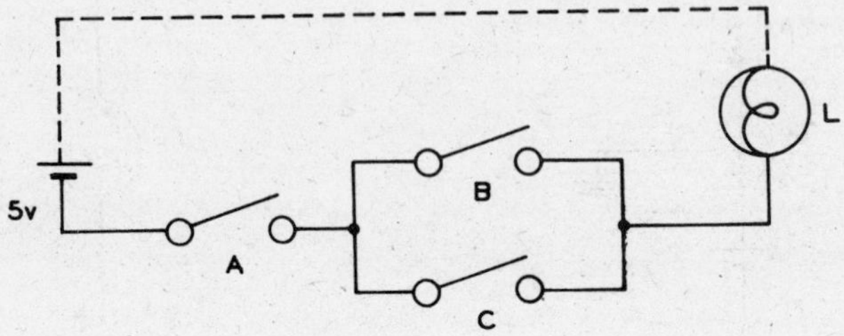


**Q:** Can you write a Boolean expression for this?

**A:** L = A • (B + C).

**Q:** Draw a truth table for this. See if the 1s in the final column agree with the conditions you determine from a physical inspection of the circuit. They should.

**A:**

| Sw A | Sw B | Sw C | Lamp |
|---|---|---|---|
| 0 | 0 | 0 | 0 |
| 0 | 0 | 1 | 0 |
| 0 | 1 | 0 | 0 |
| 0 | 1 | 1 | 0 |
| 1 | 0 | 0 | 0 |
| 1 | 0 | 1 | 1 |
| 1 | 1 | 0 | 1 |
| 1 | 1 | 1 | 1 |

If switch A and switch B are closed, then the lamp will be on. This is the condition in the seventh row of the truth table. Rows 6 and 8 can be examined in the same way.

As another example, we will use this circuit:

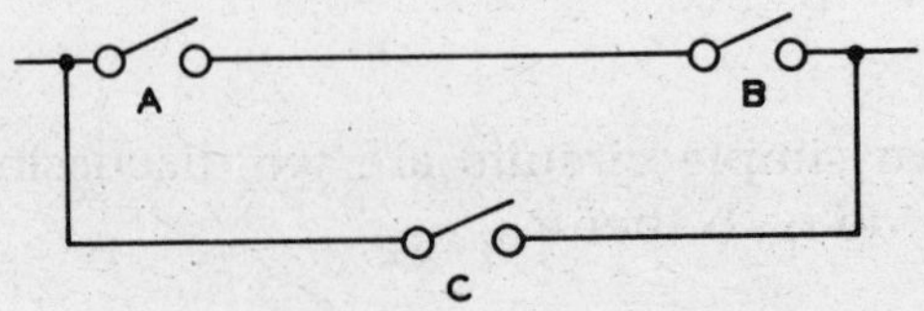

**Q:** Write the Boolean expression for this.

**A:** L = AB + C.

**Q:** Draw a truth table, and again examine it to see that it agrees with the physical conditions of the switches.

**A:**

| Sw A | Sw B | Sw C | Lamp |
|---|---|---|---|
| 0 | 0 | 0 | 0 |
| 0 | 0 | 1 | 1 |
| 0 | 1 | 0 | 0 |
| 0 | 1 | 1 | 1 |
| 1 | 0 | 0 | 0 |
| 1 | 0 | 1 | 1 |
| 1 | 1 | 0 | 1 |
| 1 | 1 | 1 | 1 |

A physical examination of the circuit will show the lamp will come on if switch C is closed. This is row 2 above. Examine the other conditions in the same way.

This is a very simple circuit:

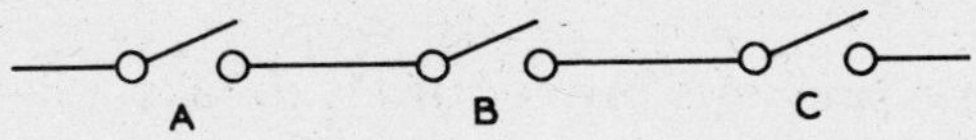


**Q:** Write the Boolean expression for this.

**A:** L = ABC.

**Q:** Draw the truth table for this.

**A:**

| Sw A | Sw B | Sw C | Lamp |
|---|---|---|---|
| 0 | 0 | 0 | 0 |
| 0 | 0 | 1 | 0 |
| 0 | 1 | 0 | 0 |
| 0 | 1 | 1 | 0 |
| 1 | 0 | 0 | 0 |
| 1 | 0 | 1 | 0 |
| 1 | 1 | 0 | 0 |
| 1 | 1 | 1 | 1 |

Finally, this simple arrangement of the switches:

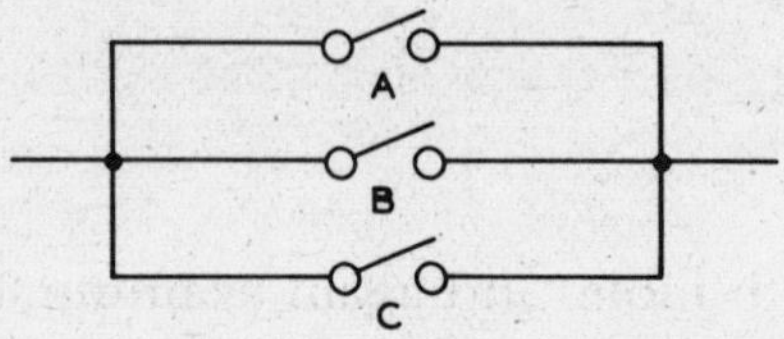


**Q:** Write the Boolean expression for this.

**A:** L = A + B + C.

**Q:** Now write the truth table.

**A:**

| Sw A | Sw B | Sw C | Lamp |
|---|---|---|---|
| 0 | 0 | 0 | 0 |
| 0 | 0 | 1 | 1 |
| 0 | 1 | 0 | 1 |
| 0 | 1 | 1 | 1 |
| 1 | 0 | 0 | 1 |
| 1 | 0 | 1 | 1 |
| 1 | 1 | 0 | 1 |
| 1 | 1 | 1 | 1 |

The reverse of the previous procedure is to take a Boolean expression and turn it into a circuit. The following examples will illustrate this easy procedure.

Consider the expression:

L = AB + BC

This represents two switches, A and B, which are ANDed together, two switches B and C which are ANDed together, and then the two ANDed combinations are ORed. If switch B is a double switch, then this circuit will be a good example.

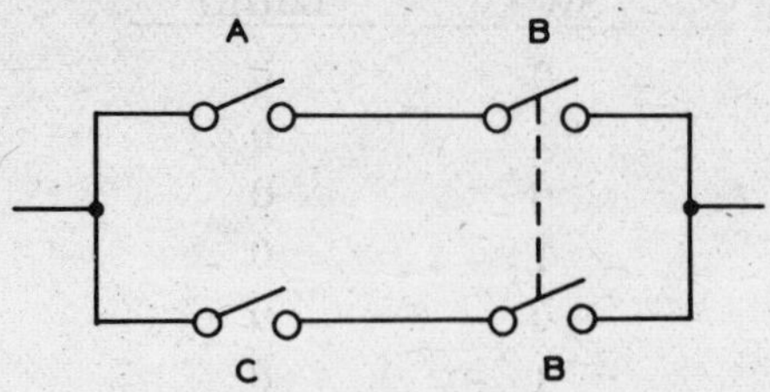


This means both poles of the switch are operated together.

However, by simplifying the expression algebraically to

$$L = (A + C) \cdot B$$

a simpler circuit can be designed.

**Q:** Draw this simpler circuit. Label the switches.

**A:**

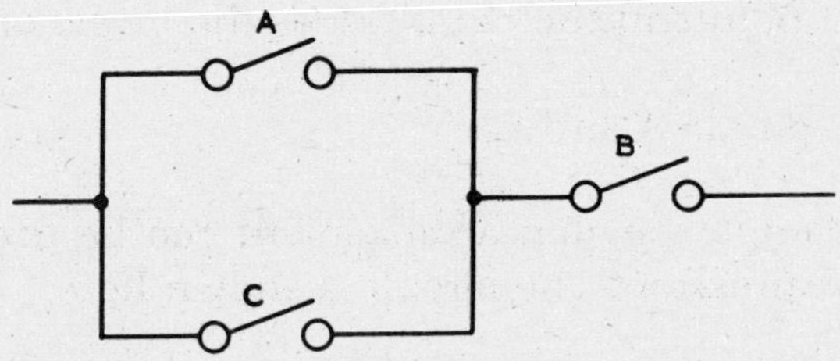


As another exercise, make this expression into a circuit:

$$L = A + BC$$

**Q:** Draw the circuit. Label the switches.

**A:**

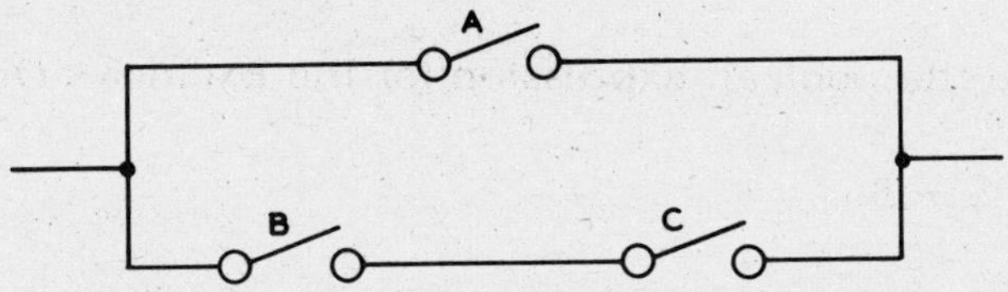


Inversion as a function is usually limited to "active" devices such as transistors. However, it can be performed mechanically with a switch and a relay, but care is needed when interpreting the meaning of the algebra.

Look at this circuit.

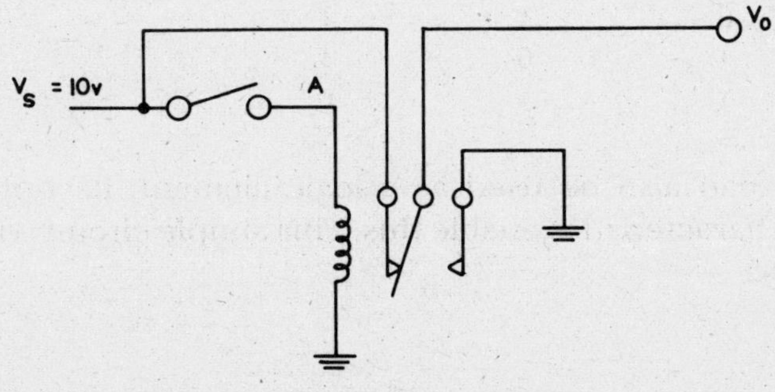

Consider the voltage levels at point A and the output voltage $V_0$.

If 0 means both the switch is open and 0V, and if 1 means both the switch is closed and +10V, then:

1. The open switch gives 0V at A and +10V out.
2. The closed switch gives +10V at A and 0V out.

**Q:** At which points in the circuit does the inversion occur?

**A:** Between points A and $V_0$.

An interesting complex switch arrangement can be used to realize the Exclusive OR expression. The circuit is shown here.

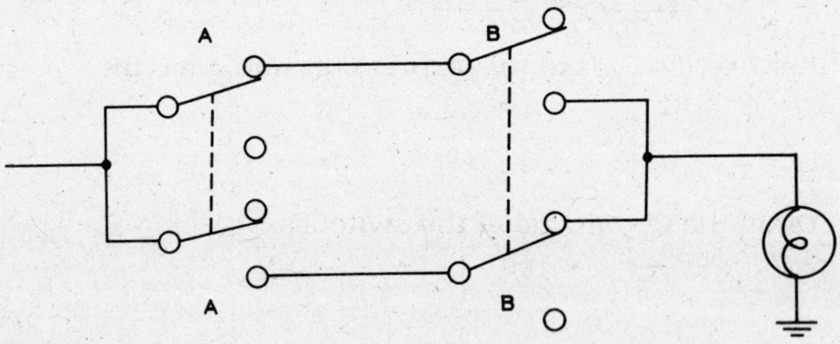


**Q:** Write the Boolean expression for the Exclusive OR.

**A:** L = A + B.

**Q:** Write a truth table for the above circuit, and check that it is an Exclusive OR.

**A:**

| Sw A | Sw B | Lamp |
|---|---|---|
| 0 | 0 | 0 |
| 0 | 1 | 1 |
| 1 | 0 | 1 |
| 1 | 1 | 0 |

The diode can also be used as a logic element. Its unidirectional conduction characteristics enable this. This simple circuit will illustrate how it is used.

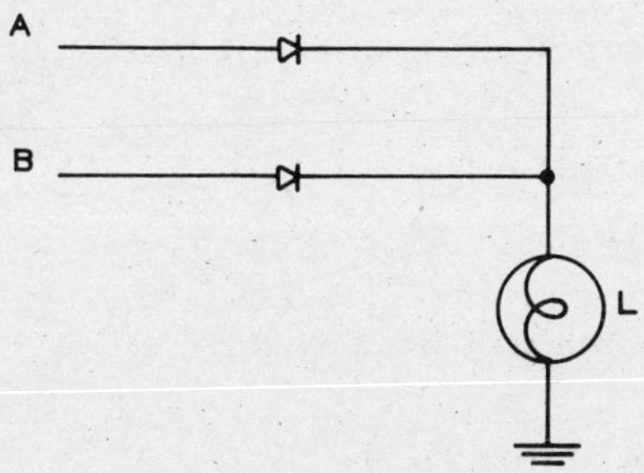


Assume that the voltages at points A and B are both low, ie. at ground level 0V. Then the voltage at point L will also be 0V.

**Q:** Is the lamp on or off?

**A:** It is off.

Now allow the voltage at point A to go high; that is, to a 1. Current can now flow through diode A and through the lamp to ground, thus completing the circuit.

**Q:** Will the lamp come on?

**A:** Yes.

Assume that the diode is perfect—that it has no voltage drop between its terminals. Then the voltage at point L will be the same as at point A. So the diode has the full-circuit current flowing through it and no voltage drop across it.

**Q:** Can you remember what other circuit component possesses this characteristic?

**A:** A closed mechanical switch.

Assume now that the voltage at A remains at 0 and the voltage at point B rises to a 1.

**Q:** Does the lamp come on?

**A:** Yes. This is really no different from the previous situation.

Assume now that both inputs A and B rise to a 1.

**Q:** What happens to the lamp?

**A:** It will again illuminate.

A truth table can be drawn for the inputs at A and B, and the condition of the lamp.

**Q:** Using 1s and 0s for the voltage levels, draw the truth table.

**A:** This is the table:

| A | B | L |
|---|---|---|
| 0 | 0 | 0 |
| 0 | 1 | 1 |
| 1 | 0 | 1 |
| 1 | 1 | 1 |

**Q:** What logic function is this table for?

**A:** The OR.

In summary: Two diodes connected as shown will perform the OR logic function. From the previous example, it may not be apparent why the diodes are necessary to illuminate the lamp. The modification shown opposite will explain this. Suppose input A has to illuminate lamp A and lamp L. Input B must illuminate lamp B and lamp L. Input A must not illuminate lamp B, nor must input B illuminate lamp A. The circuit shown on the next page will achieve this.

If input A is a 1, then lamp A will come on and so will lamp L. If input B is at 0, the diode B will be back-biased and current will not flow through it. So lamp B will be dark.

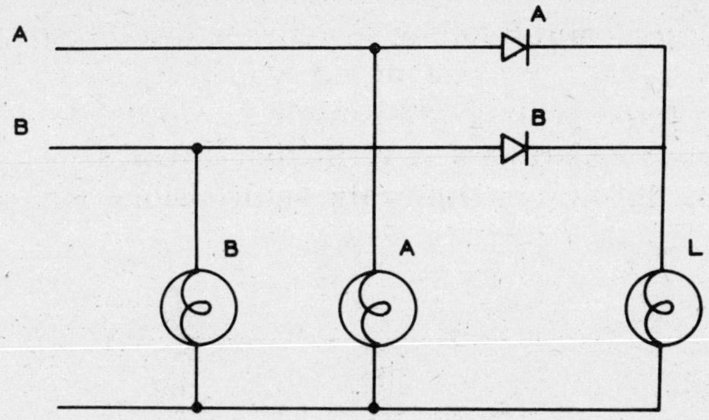


**Q:** When A is at a 1, which lamps are on and which are off?

**A:** Lamps A and L are on. Lamp B is off.

Consider what happens if the input levels are reversed. A is at 0, and B is at 1.

**Q:** Which lamps now come on?

**A:** B and L will be on. A will be off.

**Q:** Why will A be off?

**A:** Because input A is at 0, and the cathode end of diode A is at a 1. So the diode is back-biased and is not conducting current.

A truth table can be drawn for this circuit.

| Sw A | Lamp A |
|---|---|
| 0 | 0 |
| 1 | 1 |

| Sw B | Lamp B |
|---|---|
| 0 | 0 |
| 1 | 1 |

| Sw A | Sw B | Lamp A | Lamp B | Lamp L |
|---|---|---|---|---|
| 0 | 0 | 0 | 0 | 0 |
| 0 | 1 | 0 | 1 | 1 |
| 1 | 0 | 1 | 0 | 1 |
| 1 | 1 | 1 | 1 | 1 |

Note that lamps A and B follow switches A and B respectively. Note that lamp L follows the OR logic of the switches. This OR logic can be easily demonstrated with a few simple components. The student should set up this experiment to verify that it works.

The following illustration shows the setup using a few simple components.

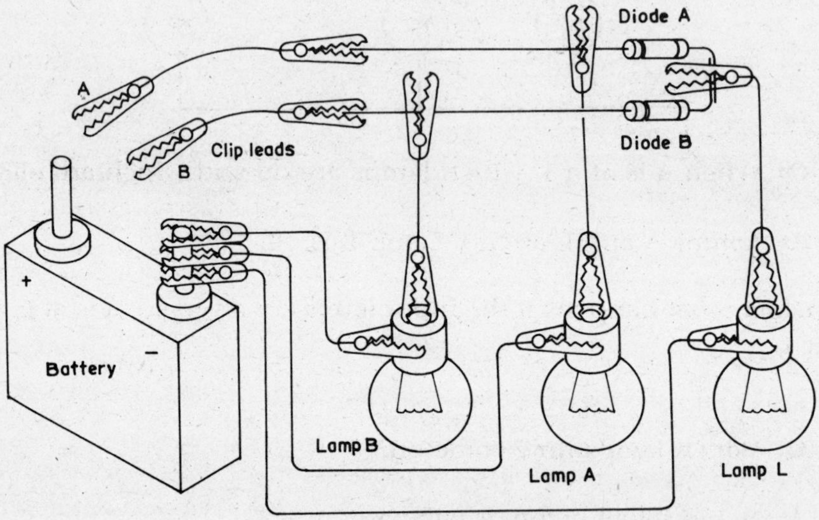


The procedure is the following:

Touch clip lead A to the positive terminal of the battery. This is presenting a logic 1 to lamp A and diode A.

**Q:** What happens?

**A:** Lamp A and lamp L come on. Lamp B stays off.

Remove clip lead A from the battery. Now touch clip lead B to the positive terminal of the battery.

**Q:** What happens now?

**A:** Lamps B and L come on. Lamp A stays off.

Now touch both clip leads to the battery.

**Q:** What happens?

**A:** All three lamps come on.

If the above actions are not observed, check these points:

1. A diode may be in the circuit the wrong way round. Ensure they are all correct.
2. Ensure all the lamps work when connected directly across the battery.
3. Ensure the battery will illuminate two lamps simultaneously.

**Q:** Draw a truth table to describe what this experiment shows.

**A:** You should get the truth table for the OR function.

We will now look at another circuit using two diodes.

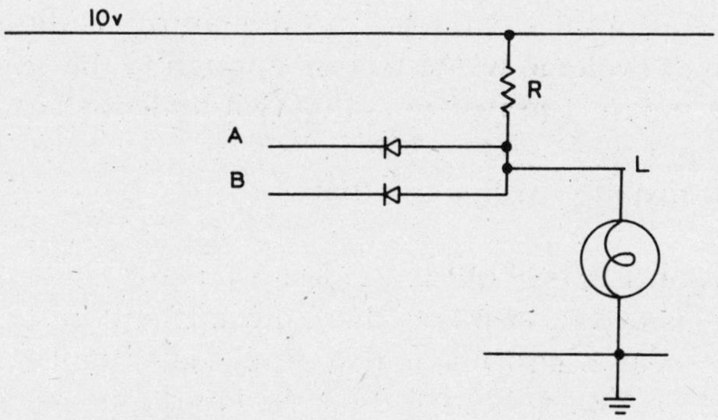


In the first place, we will assume that both A and B have low voltages; ie., that they are both at ground level or 0V. In this condition, both the diodes will be conducting.

**Q:** What will be the voltage level at point L? Assume the diodes are perfect; that is, no voltage drop across them.

**A:** Then point L will be at 0.

Now suppose point A is allowed to go to a high voltage level; that is, a 1 input occurs. Point B is still at 0.

**Q:** What is the voltage at point L now? Which diode is conducting and which is not conducting?

**A:** Point L is at 0V. This is because B is still at 0 and thus diode B is still conducting. Diode A will be reversed-biased as its cathode is higher than its anode.

Now reverse the inputs. Allow B to become a 1 and A to revert to a 0.

**Q:** What is the voltage at point L now?

**A:** Still at 0.

All three of the above conditions have kept point L at 0. Hence, the lamp has been kept off for all these conditions. Now we will allow a 1 to be present at both inputs A and B. Now there is no path for the current through the diodes to ground. Hence the current through the resistor must now flow through the lamp.

Now the voltage at point L can rise, and the lamp will illuminate. At this time, assume the voltage will rise close enough to the supply to be called a "1". This means the resistor value will be less than the resistance of the lamp.

A review of the situation will show this:

1. When A and B are both 0, the lamp L is off.
2. When A is at 1, and B is 0, the lamp L is still off.
3. When A is at 0, and B is 0, the lamp L is still off.
4. When A and B are both 1, then the lamp L is on.

**Q:** From the above information, draw a truth table for this circuit.

**A:** The truth table is this.

| A | B | L |
|---|---|---|
| 0 | 0 | 0 |
| 0 | 1 | 0 |
| 1 | 0 | 0 |
| 1 | 1 | 1 |

**Q:** What logic function is this?

**A:** The AND function.

The student should set up this circuit as an experiment and verify the AND action. Use a resistor of 100 ohms, or some close approximate value to that. The last few pages have presented both OR and AND circuits formed with mechanical switches and diodes. Although other components can be used, these two are the simplest and easiest.

Often in logic design, it does not matter which type of individual component or circuit is used; what is important is that the required logical function is performed.

To indicate which logic function is required, a symbol is used which can be written in a circuit diagram. This tells that the designer wants a particular function performed but it does not specify which components or specific circuit to use.

The symbols used are the following:

AND

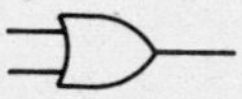
OR

# Inversion

In many instances a 1 may be present in a circuit when a 0 would be more useful, or a 0 may be present when a 1 is required. It is often necessary to change from one of these to the other.

**Q:** From the algebra, what would this action be called?

**A:** Inversion.

A circuit which performs this is called an *inverter*.

**Q:** What does an inverter do?

**A:** It changes a 1 to a 0, or a 0 to a 1.

The AND and the OR functions can be performed with mechanical switches and diodes. These are both *passive* components. Inversion usually requires an *active* component.

**Q:** Can you define "active component," and give an example?

**A:** An active component usually amplifies a current or voltage. The transistor is a good example.

In this circuit, we will represent 10V by a 1. Ground or 0V will be represented by a 0.

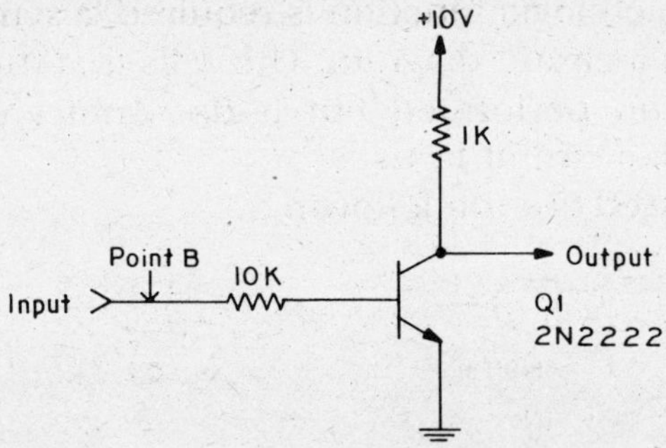


**Q:** If a 1 is put at point B, what will be the logic level at the collector?

**A:** It will be at a 0.

A truth table can be drawn for this action.

**Q:** Draw it.

**A:**

| In | Out | or | Base | Collector |
|---|---|---|---|---|
| 0 | 1 | | 0 | 1 |
| 1 | 0 | | 1 | 0 |

Obviously, these tables show a simple inversion.

In this circuit, the lamp is the collector resistor or load for the transistor, and the switch in the base circuit is used to operate the lamp and the transistor.

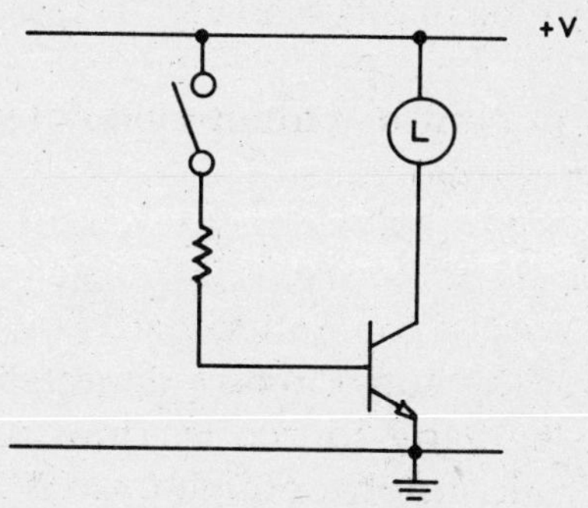


**Q:** If the switch is closed, is the lamp on or off?

**A:** The lamp is on.

In this circuit, the current through the lamp is not the same current as that through the switch. The current through the lamp is the collector current. The current through the switch is the base current. These two currents are completely isolated from each other, even though the magnitude of the base current controls the magnitude of the collector current. So, although the switch controls the state of the lamp, they are both completely isolated from each other. This isolation of the switch and the lamp is performed by the transistor.

**Q:** What is the transistor doing to the lamp and the switch?

**A:** It is isolating them from each other.

This isolation can be looked upon as a separation of the output from the input, or a separation of the devices connected to the input and output of the transistor.

**Q:** What is isolation?

**A:** Separation of input and output.

An element or component which performs this isolation is often referred to as a *buffer*. In this case, the transistor acts as a buffer between the switch and the lamp. Many elements can be used as a buffer, and buffers are quite common in electronics.

**Q:** What is a buffer in electronics?

**A:** It is a component or circuit element which isolates two or more other components.

A common use for a buffer circuit is when a source cannot supply enough voltage or current to operate a load. The source is then used to operate the buffer, and the buffer in turn operates the load. The input to the buffer does not load the source by drawing excess current. The output of the buffer will supply enough current to operate the load.

**Q:** Give a simple example of a buffer.

**A:** The transistor.

The transistor is often used as a buffer between other circuit components, but it is not the only buffer used. To indicate a buffer in digital electronics, this symbol is used.

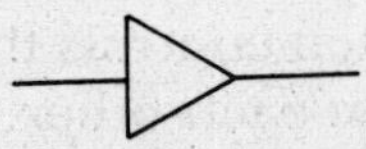

Sometimes the output of a buffer is logically the same as its input; that is, a 1 in will give a 1 out. This is called a *noninverting* buffer. The above symbol is that for a noninverting buffer.

**Q:** What is a noninverting buffer?

**A:** A buffer circuit whose output logic level is the same as its input logic level.

Some buffers are arranged to invert the logic level, so a 1 in gives a 0 out, and a 0 in gives a 1 out. This is called an *inverting* buffer. The symbol for the inverting buffer is shown here.

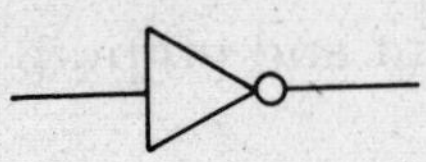

The little circle at the output means inversion. The circle is never used by itself. It is usually found on the output of an inverting buffer or some other logic symbol which performs inversion. Occasionally, it is found on an input.

**Q:** What is an inverting buffer?

**A:** A buffer which will give the opposite logic level out to that at its input.

**Q:** How is inversion indicated on a drawing symbol?

**A:** By the use of a small circle at the output. Sometimes a circle is found at the input.

# Transistor Logic

We will now take a simple diode OR gate and connect its output to a transistor inverter.

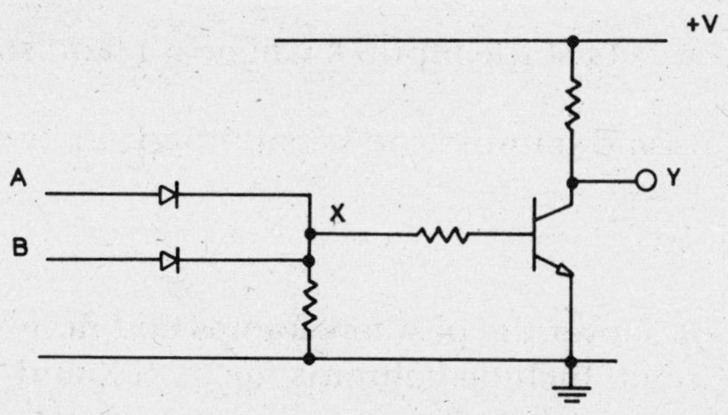


Assume that both the inputs to the diode gate are 0s.

**Q:** What is the logic level at point X?

**A:** As this is an OR gate, point X will be at 0.

**Q:** What then is the logic level at the output of the circuit; that is, at point Y?

**A:** As the transistor inverts, the output must be a 1.

So two 0s at the input will give a 1 at the output. Now allow input A to be a 1, while input B stays at 0.

**Q:** What is the level at X and at Y?

**A:** X will go to a 1, and so Y will go to a 0.

Now input A is kept at 0, and B is allowed to become a 1.

**Q:** What is the level at points X and Y?

**A:** X again becomes a 1 and Y a 0.

Now we will allow both the inputs to go to 1.

**Q:** What now are the logic levels at X and Y?

**A:** With both 1s at the input, X will be a 1, and so Y will be a 0.

This set of inputs and outputs can be summarized by drawing a truth table.

**Q:** Look back over the past few frames and draw a truth table for this circuit. Include columns for both X and Y.

**A:**

| A | B | X | Y |
|---|---|---|---|
| 0 | 0 | 0 | 1 |
| 0 | 1 | 1 | 0 |
| 1 | 0 | 1 | 0 |
| 1 | 1 | 1 | 0 |

Point X and column X are for the two diodes, which form an OR gate. Column Y is the inverse of this.

**Q:** What logic function is represented by column Y?

**A:** The NOR function.

We can conclude that an OR gate followed by an inverter will produce a NOR gate.

**Q:** Draw the circuit symbol for an OR gate, followed by the symbol for an inverter.

**A:**

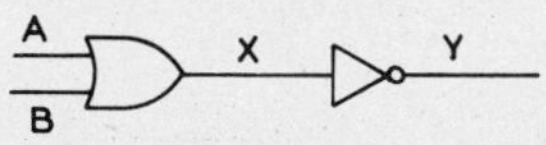


Usually a contracted form of this is used. This is the symbol for a NOR gate.

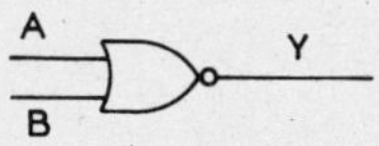


In a similar manner, a transistor can be placed at the output of a diode AND gate.

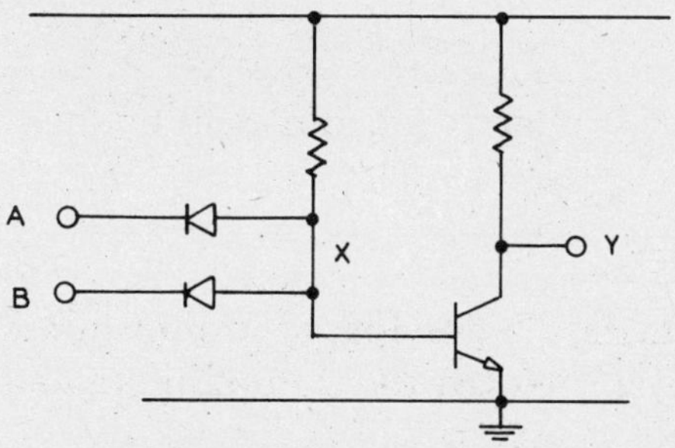


Point X will produce AND logic from the inputs. This is inverted by the transistor, and the output at point Y is the inverse of the AND.

**Q:** Draw a truth table to show this.

**A:**

| A | B | X | Y |
|---|---|---|---|
| 0 | 0 | 0 | 1 |
| 0 | 1 | 0 | 1 |
| 1 | 0 | 0 | 1 |
| 1 | 1 | 1 | 0 |

This inverse of AND is called NAND. The final column for Y is the truth table for a NAND. The NAND gate symbol is the following.

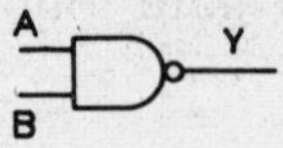


Transistors can be used by themselves to perform the logic functions without diodes, mechanical switches, or other logic components. We will examine this circuit.

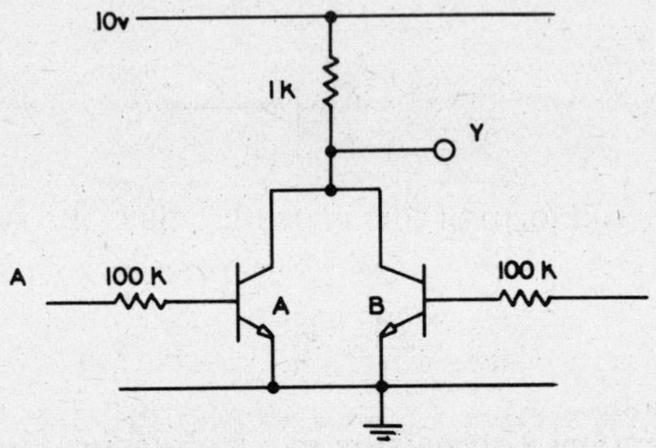


Suppose both inputs are at 0. In this condition, no base current will be flowing in either transistor.

**Q:** How much collector current will be flowing?

**A:** None.

If no collector current is flowing, then both transistors are OFF, and no current is flowing through the common collector load resistor.

**Q:** What will be the voltage output at point Y?

**A:** It will be high, at 10V. This is a logic 1.

So two 0s at the input will give a 1 at the output. Now allow input A to be a 1 and B to remain at 0.

**Q:** Does any base current flow now?

**A:** Yes, in transistor A.

**Q:** Does any collector current now flow?

**A:** Yes, in transistor A.

As this collector current is flowing in transistor A, it must also flow through the common load resistor. So this load resistor has a voltage drop across it. From the values shown in the figure opposite, the base current, collector current, and the voltage drop across R can be calculated.

**Q:** Using the values shown in the figure above, calculate 1) the base current; 2) the collector current; and 3) the voltage drop across R.

**A:** 1) $I_B = V/R_B = 10V/100K = 0.1$ ma
2) $I_C = I_B \bullet B = 0.1 \text{ ma} \times 100 = 10$ ma
3) $V_R = R \bullet I_C = 1K \times 10 \text{ ma} = 10V$

If the voltage drop across the common load resistor is 10V, we can easily find the voltage output at point Y.

**Q:** What is it?

**A:** 0V, which is a logic 0.

Suppose we now repeat this with input A at 0 and with input B at a logic 1.

**Q:** Is this the same as before?

**A:** Yes.

**Q:** So what then is the output logic level?

**A:** 0 again.

From this we can conclude that a 1 in will give a 0 out. Obviously, if we have 1s at both inputs, then we will still get a 0 output. Now summarize the inputs and corresponding outputs in a truth table for this circuit.

**Q:** Draw up the table.

**A:**

| A | B | Y |
|---|---|---|
| 0 | 0 | 1 |
| 0 | 1 | 0 |
| 1 | 0 | 0 |
| 1 | 1 | 0 |

From our previous knowledge, we know what logic function this table represents.

**Q:** What is the logic function?

**A:** The NOR function.

Here the inherent inversion characteristic of the transistor has also performed a logic operation on the input signals, and the NOR has been produced instead of a simple OR.

An addition of another transistor to this circuit, as shown here, will perform another inversion on the output.

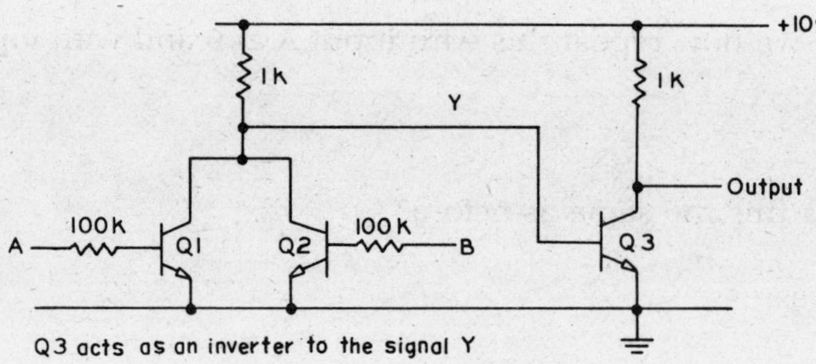


Q3 acts as an inverter to the signal Y

**Q:** What will be the output now, which logic function?

**A:** It will be an inverted NOR, which is an OR.

From a superficial examination of the above OR and NOR circuits, it is easy to see that the NOR is simpler.

**Q:** Why is the NOR simpler?

**A:** Because an extra transistor is needed for the OR.

Suppose we now hook up two transistors in this way.

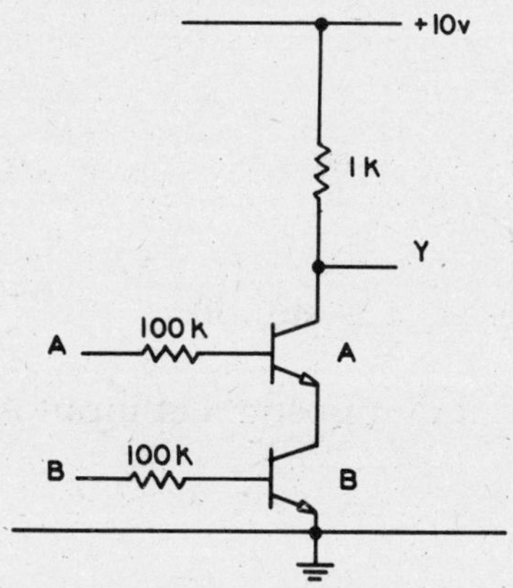


Assume that both inputs A and B are at 0. In this case, no base current will be flowing. And so there will be no collector current. Both transistors are off; they are like open switches.

**Q:** What is the voltage at point Y?

**A:** It is a 10V. A logic 1.

So here two 0 inputs will produce a 1 output. Suppose now input A becomes a 1, while B remains at 0. Base current will try to flow through the path shown, on the next page.

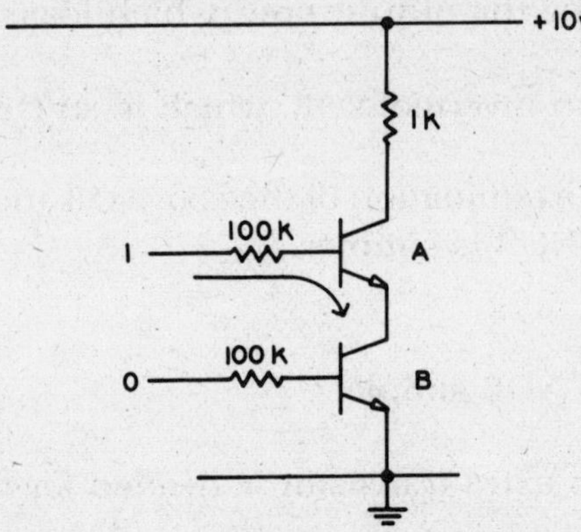


It cannot flow as there is no complete path or complete circuit for it to flow. The reason is that transistor B is still off. It is conducting no current.

**Q:** When input A goes to a 1, how much base current flows in transistor A? Why?

**A:** None. Transistor B is still off.

Suppose now we do the opposite. Let input A remain at 0 and let B rise to a 1.

**Q:** Will base current flow in transistor B? Why?

**A:** Yes, it will flow. There is a complete circuit for it to flow around. See this figure.

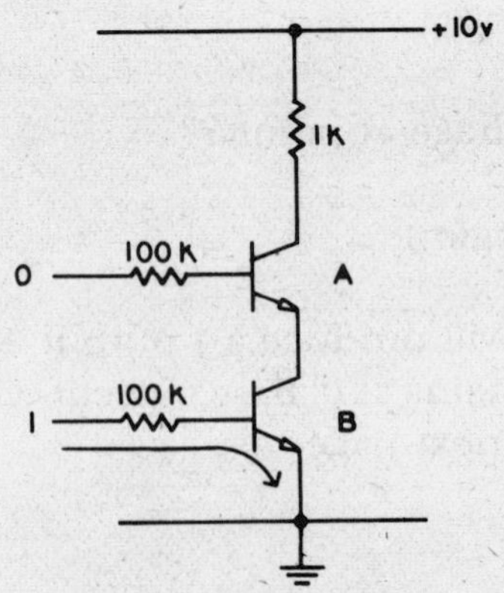

Now although base current is flowing in transistor B, there is no collector current flowing. This is because transistor A is off; it is like an open switch.

**Q:** When input B goes to a 1 and A remains at 0, how much collector current flows in B? Why?

**A:** None. Transistor A is off and no current can flow through it.

So, summarizing these last two situations, if any of the inputs goes to a 1 while the other remains at a 0, then no collector current will flow through the transistors.

**Q:** In that case, what will be the output voltage?

**A:** It will remain at 1.

Now for the final situation. Let both inputs rise to a 1.

**Q:** What happens to transistor B?

**A:** It turns on.

As transistor B is on, it provides a complete path for the base current for transistor A. Look at this drawing.

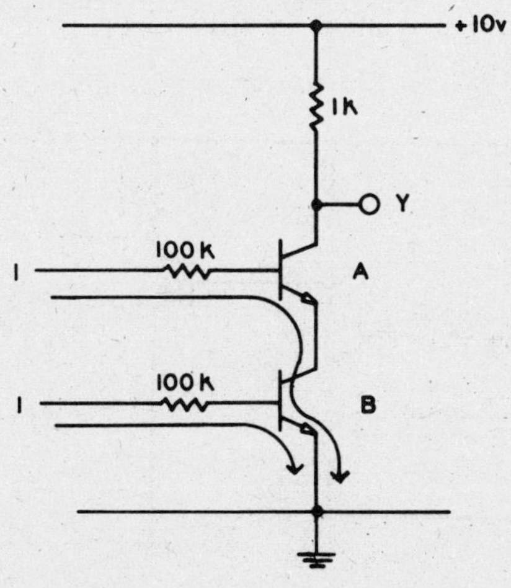

As transistor A now has base current flowing, it can also have collector current flowing. This collector current will now flow through transistor B as it is on. So now current will flow through the load resistor.

**Q:** What happens to the voltage at point Y?

**A:** It will drop.

Correct choice of resistor values will ensure the current through the load resistor will bring the voltage at Y down to a logic 0. So two inputs at a logic 1 will produce a logic 0 output.

**Q:** Summarize the inputs and outputs in a truth table.

**A:**

| A | B | Y |
|---|---|---|
| 0 | 0 | 1 |
| 0 | 1 | 1 |
| 1 | 0 | 1 |
| 1 | 1 | 0 |

**Q:** From previous theory, what is the logic function represented by this table?

**A:** The NAND function.

By placing an extra transistor in this circuit, as shown here, the output can be inverted.

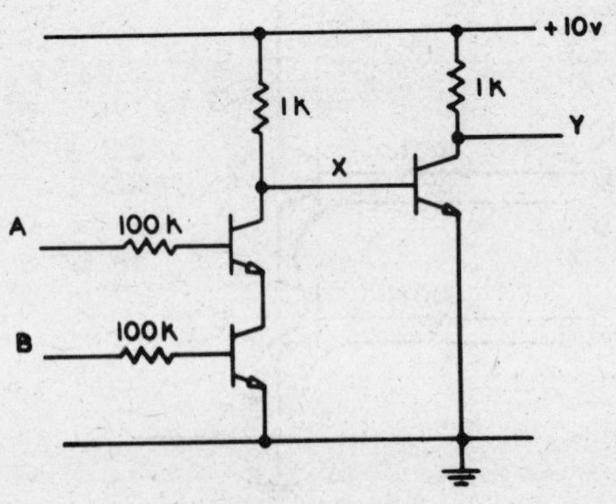

**Q:** Draw a truth table for this new output Y.

**A:**

| A | B | X | Y |
|---|---|---|---|
| 0 | 0 | 1 | 0 |
| 0 | 1 | 1 | 0 |
| 1 | 0 | 1 | 0 |
| 1 | 1 | 0 | 1 |

**Q:** What logic function does this last column represent?

**A:** The AND function.

So, as with the OR and NOR, the inversion characteristic of the transistor will produce a NAND easier than it will produce an AND. An AND gate followed by an inverter can be shown symbolically in this form. This is a NAND gate.

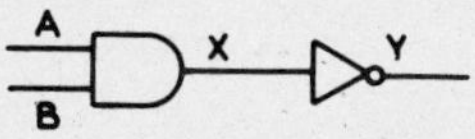


Usually a NAND gate is indicated by this symbol.

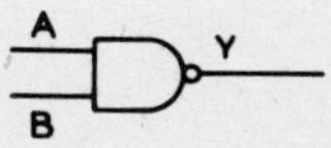


Look at the NAND gate shown in this figure.

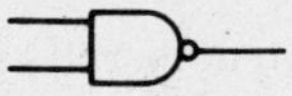

Suppose both inputs were connected together and only one input signal applied.

**Q:** Draw a truth table for this connection.

**A:**

| In | Out |
|---|---|
| 0 | 1 |
| 1 | 0 |

**Q:** What function does this represent?

**A:** A simple inversion.

We will repeat this for the NOR gate.

Both inputs are connected together and only one input signal is applied.

**Q:** Draw the truth table.

**A:**

| In | Out |
|---|---|
| 0 | 1 |
| 1 | 0 |

**Q:** What is this?

**A:** Inversion again.

From this we see that both the NAND and the NOR gates can be used as inverters. Later we will see that ALL logic functions can be performed with combinations of the NAND and the NOR gates only.

In fact, all logic functions can be performed solely with NAND gates. This is very important in modern electronics due to the fact that these type gates are easier to manufacture.

As an example of this, we will consider De Morgan's Laws. The first De Morgan Law is this:

$$\overline{A + B} = \overline{A} \cdot \overline{B}$$

NOR AND

The LHS is the NOR of two positive propositions. The RHS is the AND of two negative propositions; in fact, they are the inverses of the positive propositions on the LHS.

**Q:** Try to draw out the simplest logic circuits for each side of the expression.

**A:** The simplest circuits would be these.

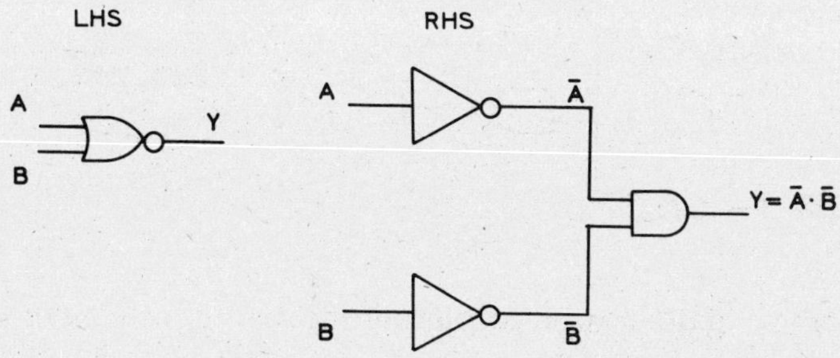


Obviously, the LHS is much simpler than the RHS. A large part of Boolean algebra use in digital electronics is to see where De Morgan's Laws can be used to effect such circuit simplifications. De Morgan's Second Law is this:

$$\overline{A \cdot B} = \bar{A} + \bar{B}$$

NAND OR

The LHS is the NAND of two positive propositions. The RHS is the OR of two negative propositions.

**Q:** Try to draw out the simplest logic circuits for each side of the expression.

**A:** These are the simplest.

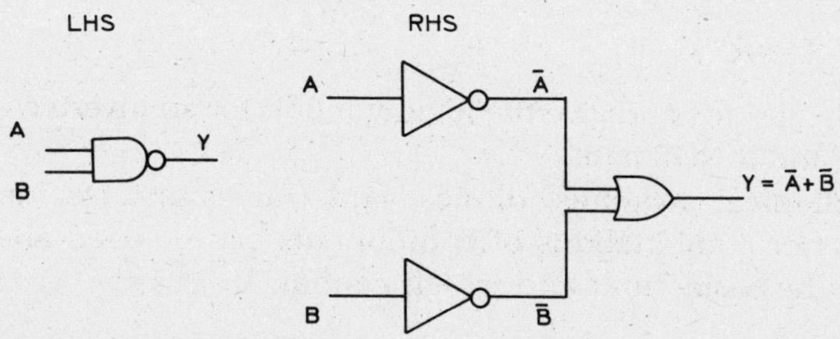


Again, one is much simpler than the other. So this law can also be used for circuit simplification.

An interesting exercise is to implement the first law using only NAND gates. Remember how to use the NAND as an inverter.

**Q:** Draw a circuit for a NOR gate using only NAND gates

**A:** This is a composite NOR gate made from NAND gates.

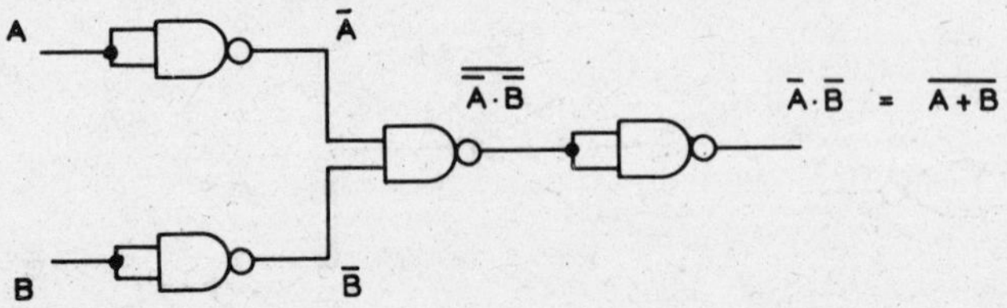


Similarly, a NAND gate can be implemented using only NOR gates.

**Q:** Do this.

**A:** This is a NAND gate made from NOR gates.

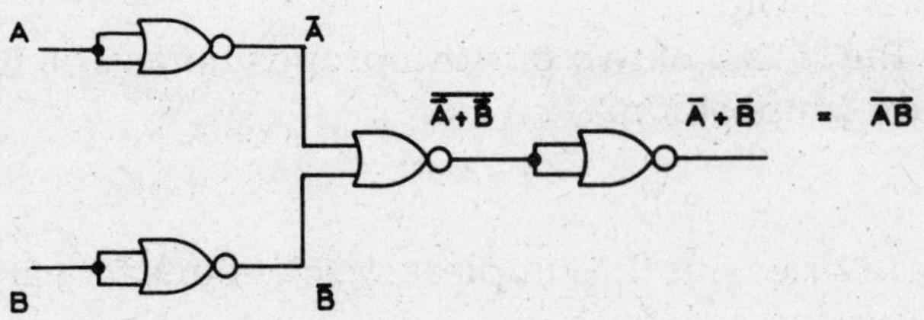


## Self-Test

Draw simple circuits and give the logic symbols for an inverter, an AND, OR, NAND, and a NOR gate.

Use mechanical switches, diodes, and transistors. Decide which components or combinations of components can be used and which circuits can be easily built and which cannot.

## Answers

This table shows those circuits which are reasonable by using a 1. Circuits which are unreasonable are indicated by a 0. The actual circuits can be found in the text of the chapter.

| | INV | AND | OR | NAND | NOR |
|---|---|---|---|---|---|
| Mechanical Switches | 0 | 1 | 1 | 0 | 0 |
| Diodes | 0 | 1 | 1 | 1 | 1 |
| Diodes and Transistors | 0 | 1 | 1 | 1 | 1 |
| Transistors only | 1 | 1 | 1 | 1 | 1 |

CHAPTER
# 6

# Flip-Flops

**OBJECTIVES:** Upon completion of Chapter 6 you will be able to do the following:

- ☐ Describe the main characteristics of flip-flop circuits
- ☐ Identify various types of flip-flop circuits
- ☐ Work experiments using various flip-flop circuits
- ☐ Identify waveforms and pulses associated with digital circuits
- ☐ Observe digital pulses with an oscilloscope
- ☐ Calculate period and frequency of various waveforms

ONCE INFORMATION HAS BEEN ENCODED or generated in binary or digital form, it is then usually necessary to process it in some way. For example, in a computer the information may be received in *serial* form, one bit following another until it is all in the computer. But it may be necessary to work on all the bits simultaneously; that is, in *parallel* form.

This occurs when entering a number from a keyboard. The digits are generated successively but the computer must have them all at once to perform addition. To be able to do this, the computer must be able to store the information and to convert it from serial to parallel form.

To store information, some sort of memory is used. One way of storing information electronically is to use a circuit called a *flip-flop*. There is no single or unique circuit for a flip-flop; many different ones have been devised. Some of the simpler and more basic ones are shown in this chapter, using both transistors and logic blocks.

In this section, the student will see:

1. The basic inverter flip-flop
2. The NAND flip-flop
3. The NOR flip-flop
4. The RS flip-flop
5. The "clocked" RF flip-flop
6. The "T", the "D", and the "JK" flip-flop

Semiconductor memories are covered in Chapter 7, but they are all one form or another of a flip-flop.

Before starting on this section, the student should be satisfied that he or she knows the material on transistor switches. Many of the questions and answers in this section rely heavily on knowledge gained in the transistor switching chapter.

Consider a transistor setup as in this figure.

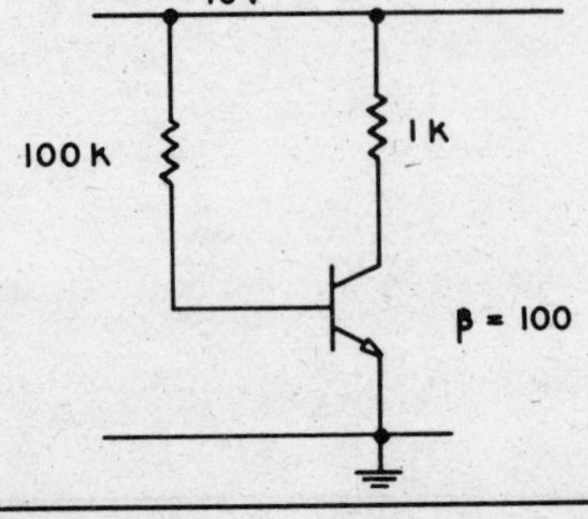

**Q:** From the values given, is it ON or OFF? You have worked this out before.

**A:** It is ON.

**Q:** What is the voltage at the collector?

**A:** 0V.

Suppose this transistor is now connected to an identical transistor, as shown here.

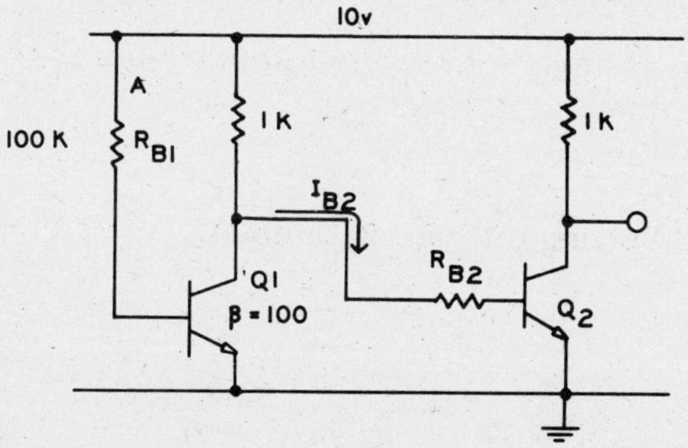


**Q:** How much base current $I_{B2}$ is flowing? Why?

**A:** None. The left-hand end of $R_{B2}$ is at 0V.

**Q:** Is $Q_2$ ON or OFF? Why?

**A:** OFF. There is no collector current flowing.

Now, holding this situation in mind—$Q_1$ ON and $Q_2$ OFF—look at the end of $R_{B1}$ labelled A.

**Q:** What is its voltage?

**A:** 10V.

Let us now move this end of $R_{B1}$ to the collector of $Q_2$. This is shown in the figure at the top of the following page.

**Q:** What is the voltage at this end of $R_{B1}$? Has it changed?

**A:** It has not changed; it is still 10V.

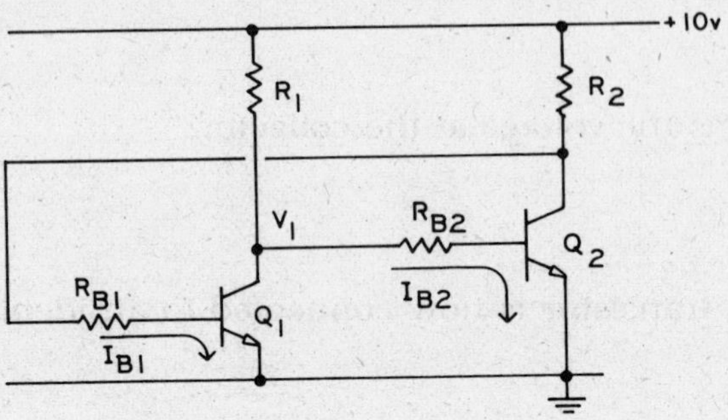


Having made this change in the circuit, think about these questions before you answer them.

**Q:** Is $I_{B1}$ still flowing through $R_{B1}$ into $Q_1$?

**A:** Yes.

**Q:** Is $Q_1$ still ON?

**A:** Yes.

**Q:** What is $V_1$?

**A:** It is still 0V.

**Q:** Is $Q_2$ still OFF?

**A:** Yes.

**Q:** Is the collector of $Q_2$ still at 10V?

**A:** Yes.

So we have come full circle, back to where we started, and found that the circuit has not changed at all. Does it seem likely that anything will happen in this circuit? In fact, nothing will happen. It will remain in this condition all the while the power is turned on. It is in a stable position.

In this figure, the circuit is redrawn to emphasize the symmetrical nature of the circuit. This is the usual way this circuit is drawn.

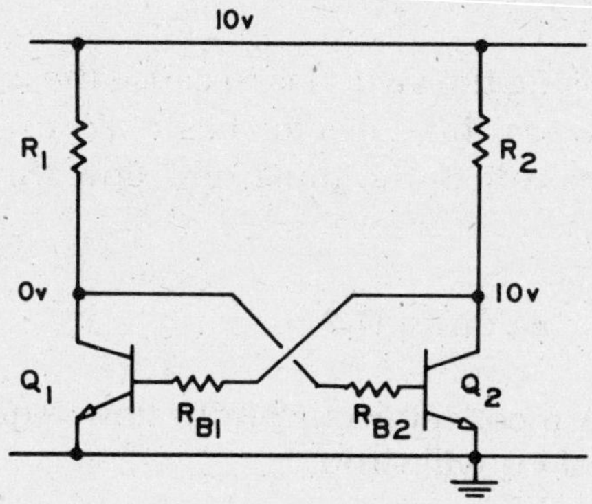


Now let us see what can be done with the circuit. It should be obvious that all we can do is to turn $Q_1$ OFF and $Q_2$ ON, in other words, to change the state of both of them. Suppose the base of $Q_1$ is shorted to ground with the switch.

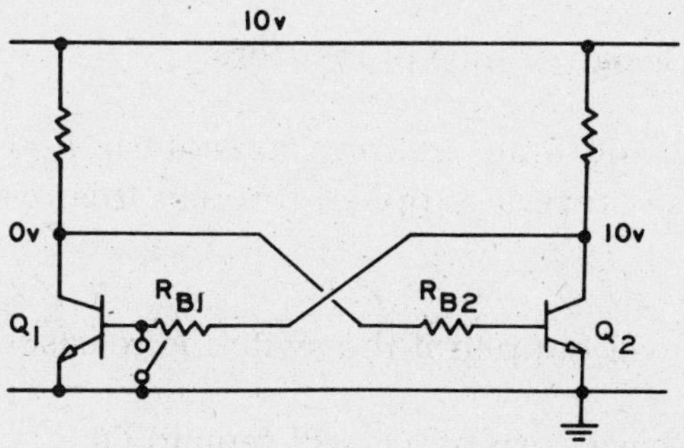


When the switch is closed, then the base voltage is tied to ground.

**Q:** How much base current now flows into $Q_1$?

**A:** None; it all now flows to ground through the closed switch.

**Q:** How much collector current is now flowing in $Q_1$?

**A:** None.

**Q:** So what must the voltage at the collector of $Q_1$ be?

**A:** It must be 10V.

So the effect of closing the switch is to cause the collector of $Q_1$ to rise from 0V to 10V. Now, as the collector has risen to 10V, then $R_{B2}$ must have a 10V drop across it. So $I_{B2}$ must now flow through it.

**Q:** What will this do to $Q_2$?

**A:** It will cause collector current to flow. And the voltage at the collector of $Q_2$ will drop.

If the resistor values have been properly chosen, then the collector voltage of $Q_2$ will drop to 0V.

**Q:** How do we ensure that $Q_2$ collector will drop to 0V?

**A:** Correct design choice of resistors.

With $Q_2$ collector now at 0V, then no current can flow through $R_{B1}$. So $Q_1$ now has no base current supplied through this resistor.

**Q:** So what will happen if the switch is released?

**A:** With no base current, $Q_1$ will remain OFF.

Let us review the situation. Pressing and then releasing the switch has caused $Q_1$ to go from ON to OFF, and $Q_2$ has gone from OFF to ON. They have both changed states.

**Q:** Draw the circuit with the new voltage values at the collectors.

**A:**

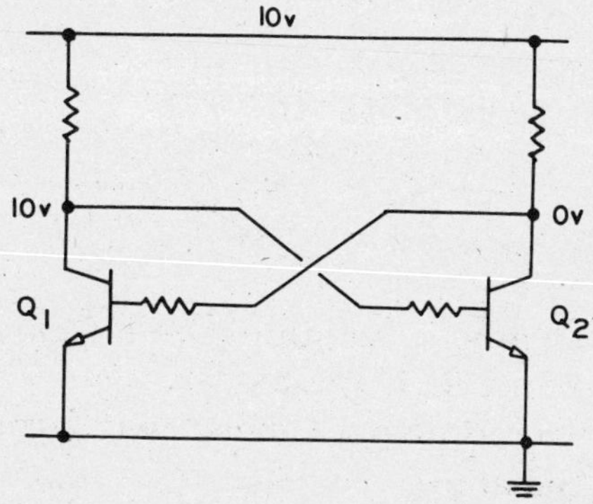


What happens if the switch is operated again? As the base of $Q_1$ is at ground and no base current is flowing, the switch will not do anything to alter this. So another operation of the switch will have no effect. The momentary closure of the switch has been "remembered" by the circuit.

It is "remembered" as the circuit will stay in the condition into which the switch closure put it, and it will stay there after the switch reopens. As the static state of the two transistors has changed over, the circuit is said to have "flipped" or "flopped" to its other state. This is why it is called a flip-flop.

**Q:** From the previous frames, what is the main characteristic of a flip-flop circuit?

**A:** It has two stable states.

At this point, the reader should realize that the flip-flop has many uses. Instead of writing out flip-flop, the two letters FF are used. The FF is one of a class of circuits called *multivibrators*. We will see more of these later. As the FF is stable in both states, it is also called a *bistable* multivibrator.

**Q:** What is another name for a bistable multivibrator?

**A:** A flip-flop, or an FF.

The word multivibrator is also shortened. MV is used.

The two single transistors used here are both inverters. So the circuit can be drawn with the logic symbols.

Note that with logic symbols the supply voltage and ground are omitted. The only reason for this is visual clarity.

This circuit can be redrawn in these two forms, which are more usually found than the above.

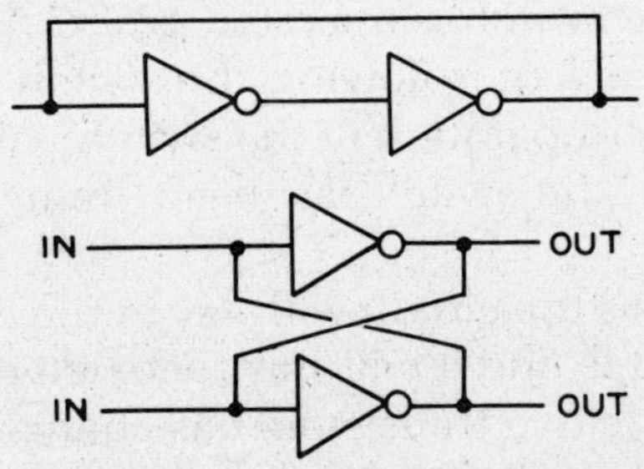


**Q:** Why has no supply voltage or ground been shown in these?

**A:** Simply to make the drawing easier to see.

Note the inputs and outputs have been labelled. A simple FF of this nature is seldom used. A common refinement is to use NAND gates instead of inverters, with the advantage of separating the inputs from the outputs.

# Flip-Flop Experiment

Set up the circuit shown in the figure on the top of page 167. Alternately, operate the buttons and measure the voltages at the two collectors after each operation. The voltages should always be either 0V or 10V, and the collectors should never be at the same voltage simultaneously.

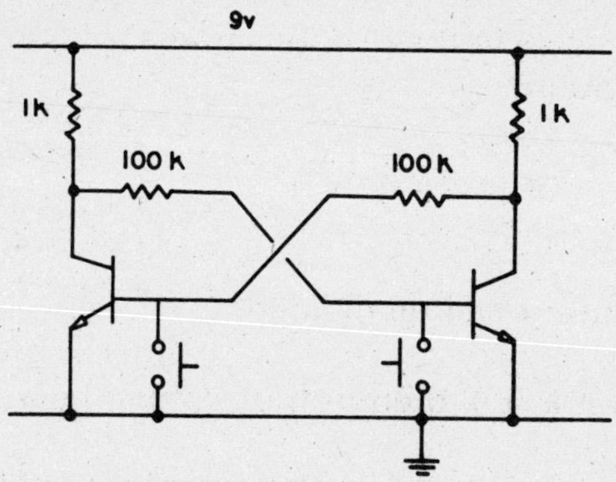


Any NPN transistors can be used. The 2N2222 is a good choice.
We will start this section with a question.

**Q:** Draw a NAND symbol and the truth table for this function.

**A:**

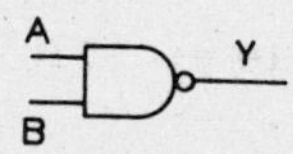


| A | B | Y |
|---|---|---|
| 0 | 0 | 1 |
| 0 | 1 | 1 |
| 1 | 0 | 1 |
| 1 | 1 | 0 |

Now, look at the two NAND gates connected in this fashion.

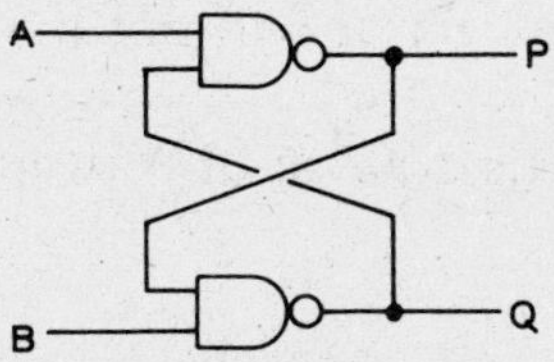


Assume that input A is at a 0, and input B is at a 1. Can we determine what the outputs will be? With a 1 input, a NAND gate can have either a 1 or a 0 output. A 1 input does not uniquely set or define the output. With a 0 input, a NAND gate must have a 1 output. Look at the truth table and satisfy yourself of this.

**Q:** From the facts in the previous frame, can an output for either gate be found?

**A:** Yes.

**Q:** Which output can be defined?

**A:** As input A is a 0, then output P must be a 1.

Now, this output P is fed back to the other input of gate B. Gate B has two inputs which are 1s.

**Q:** From the truth table, what is the output of gate B?

**A:** Output Q is a 0.

Now, we will allow input A to go to a 1, and we will see if this affects the outputs.

**Q:** If input A goes to a 1, what are the two inputs to gate A?

**A:** A 1 and a 0.

**Q:** For a 1 and 0 input to a NAND gate, what is its output?

**A:** Output P will be a 1. It will not have changed.

Suppose now we let this 1 at input A drop back to a 0.

**Q:** Does this change anything?

**A:** No, output P still remains a 1.

Note that as we did not change input B, and P remained a 1; then nothing happened to alter output Q.

Let us summarize these changes at the input A with a truth table.

| A | B | P | Q |
|---|---|---|---|
| 0 | 1 | 1 | 0 |
| 1 | 1 | 1 | 0 |
| 0 | 1 | 1 | 0 |

Obviously, if we started with the inputs reversed so that A was a 1 and B a 0 and then changed input B, the results would be similar.

**Q:** Draw a truth table for this set of conditions.

**A:** It will be this.

| A | B | P | Q |
|---|---|---|---|
| 1 | 0 | 0 | 1 |
| 1 | 1 | 0 | 1 |
| 1 | 0 | 0 | 1 |

Note that columns P and Q are reversed from the above. Suppose now that both inputs A and B are set at 0, what are the outputs?

**Q:** Look at the truth table for the NAND and determine the outputs.

**A:** They are both 1s.

For various reasons, this condition is avoided in digital electronics. Having both outputs of a FF the same is not usually desired.

The interesting action in this FF circuit occurs when both inputs are the opposite of each other, and when they both change simultaneously.

Let us start with a 0 at A and a 1 at B.

**Q:** Draw a one-row truth table for this.

**A:**

| A | B | P | Q |
|---|---|---|---|
| 0 | 1 | 1 | 0 |

If the inputs now change simultaneously, then the 0 input at B will force a 1 to appear at Q. Gate A will now have both inputs at 1.

**Q:** From the truth table, what will its output be?

**A:** This is the only condition where the output can be 0.

So now we can draw a one-line truth table for this condition.

**Q:** Draw it.

**A:**

| A | B | P | Q |
|---|---|---|---|
| 1 | 0 | 0 | 1 |

A complete table for this FF can now be drawn.

| A | B | P | Q |
|---|---|---|---|
| 0 | 1 | 1 | 0 |
| 1 | 0 | 0 | 1 |

We have previously seen that two 1 inputs will cause no change, no matter what state the FF is in.

This circuit is called an *R-S* flip-flop. The inputs we called A and B are usually called R and S. These stand for reset and set. It is shown below with another common symbol.

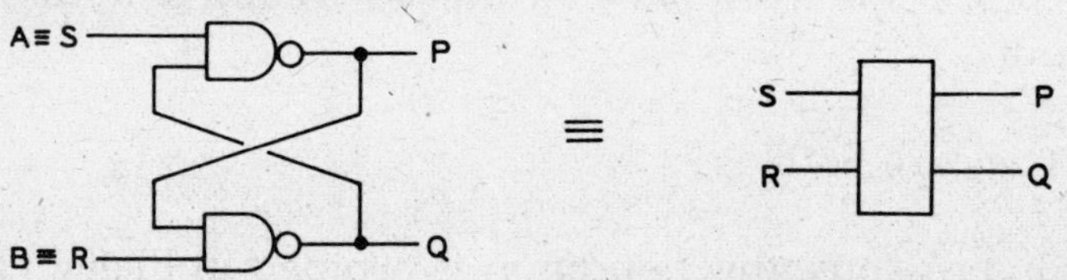


The condition where $Q = 1$ and $\overline{Q} = 0$ is called the *SET* condition. The opposite where $Q = 0$ and $\overline{Q} = 1$ is called *RESET*.

A 1 at input B causes the RESET condition, so this is called the *R input*. A 1 at input A causes the SET condition, so this is called the *S input*. The RS FF is the basic FF, and all others are refinements of this. The main characteristic of FFs is that the two outputs are ALWAYS in opposite logic states.

**Q:** If the Q output is 1, what is the $\overline{Q}$ output?

**A:** It will always be a 0.

The RS FF can also be made with NOR gates.

**Q:** Draw a NOR gate symbol and write out its truth table.

**A:**

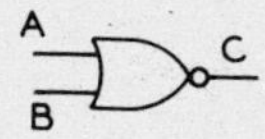


| A | B | Y |
|---|---|---|
| 0 | 0 | 1 |
| 0 | 1 | 0 |
| 1 | 0 | 0 |
| 1 | 1 | 0 |

Two NOR gates, cross-coupled in this way, will form an RS FF.

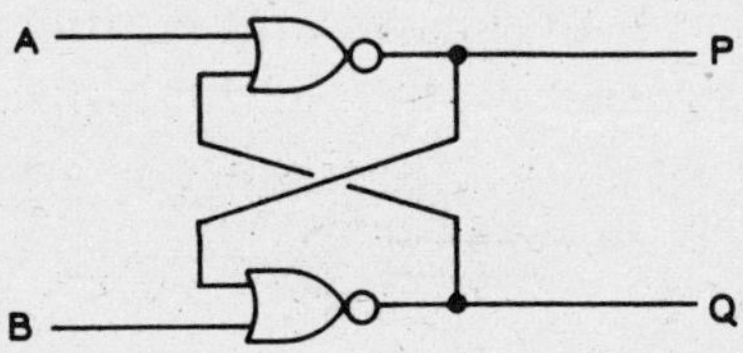


This can be examined by starting with a 0 at A and a 1 at B.

**Q:** With these inputs, can any output be uniquely determined?

**A:** From the truth table, a 1 in will always give a 0 out. So Q will be at 0.

If the inputs were reversed, then the outputs would also be reversed.

**Q:** Draw a truth table for both situations.

**A:**

| A | B | P | Q |
|---|---|---|---|
| 0 | 1 | 1 | 0 |
| 1 | 0 | 0 | 1 |

If both inputs are allowed to be 1s, then both outputs will be 0s. This is a not-allowed condition and so this input combination is avoided. With both inputs at 0, the output cannot be uniquely determined.

**Q:** Write out the complete truth table; show all four input conditions and label all four output conditions.

**A:**

| A | B | P | Q | |
|---|---|---|---|---|
| 0 | 0 | Indeterminate | | |
| 0 | 1 | 1 | 0 | |
| 1 | 0 | 0 | 1 | |
| 1 | 1 | 0 | 0 | This is not allowed. |

As with the NAND gate RS FF, both inputs are made to change simultaneously to avoid the unwanted conditions.

How are we to ensure that the inputs to the FFs just discussed will change simultaneously when we cannot guarantee that the outputs driving them will be so obliging? This next modification explains how.

Suppose we take two NAND gates and connect them as shown here. Note that there is no cross-coupling of the outputs back to the inputs.

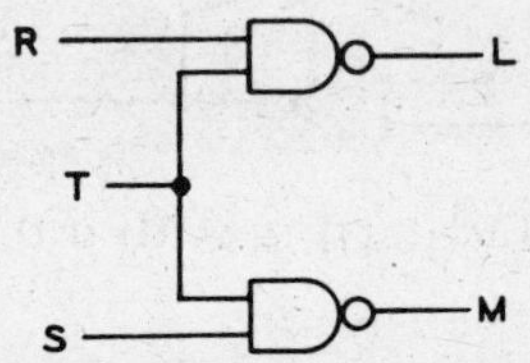


Note that this circuit is NOT a FF. Let us see if we can determine what it does.

**Q:** To get started, draw the table for a NAND gate.

**A:** A NAND truth table:

| A | B | Y |
|---|---|---|
| 0 | 0 | 1 |
| 0 | 1 | 1 |
| 1 | 0 | 1 |
| 1 | 1 | 0 |

Now, back to the circuit, we will start with all three inputs at logic 0.

**Q:** What will be the outputs L and M?

**A:** Both will be at 1.

Let R become a 1, and S and T remain at 0.

**Q:** What are L and M?

**A:** The 0 on T ensures L stays at 1. Both the zeros on S and T will keep M at 1.

If we reverse these conditions and let S become 1 instead of R, while R and T stay at 0, what then happens? Obviously, nothing. L and M still stay at 1. What happens now if we let T become 1? Assume that R and S remain at 0.

**Q:** What are L and M now?

**A:** Again, they stay at 1.

Suppose now we have this condition:

R = 1, S = 0, and T = 1

**Q:** Work out L and M now.

**A:** R = 1 and T = 1 will give a 0 at L. The 0 at S will keep M = 1.

**Q:** Do a truth table line for this.

**A:**

| T | R | S | L | M |
|---|---|---|---|---|
| 1 | 1 | 0 | 0 | 1 |

The reason a change in the outputs was able to occur is that T has gone to a 1. When it was at 0, it forced both L and M to be 1s. So inputs R and S can only pass through the gates to affect L and M when T is at 1.

**Q:** What happens when T returns to 0?

**A:** Then both L and M will revert to 1s.

Note that L will be the inverse of R when T = 1 and that M will be the inverse of S when T = 1. So all the while T = 1, both R and S can pass through the gates in their inverted form.

When T = 0, the outputs of the gates are locked into the logic 1 level. In this condition, they are said to be *inhibited*.

**Q:** What is an inhibited gate?

**A:** One whose output is locked to a definite logic level, and whose inputs will not affect the output.

These gates are enabled or *deinhibited* only when the T input goes high. The T input is thus called a *toggle* input. Hence, the use of T. The T input is sometimes called a *gating* input.

Often the input to the T is a series of regularly timed, positive going pulses.

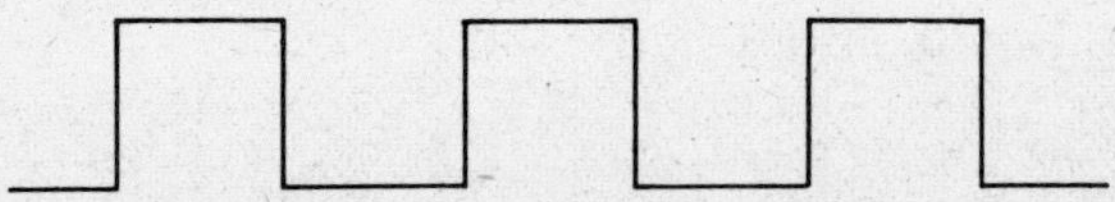

These appear as regularly as clockwork, so they are called *clock pulses*. T is thus called a *clock* input. The value of these two gates with their clock input lies in the fact that both outputs are at 1 when the clock or T input is at 0.

Look back at the truth tables for the NAND RS FF.

**Q:** What is the importance of the output condition when both inputs are 1s?

**A:** If both inputs are 1s, they do not cause any change to the output as it was set up by the previous input conditions.

Let us now set up this circuit. The two gates are hooked to the front of a NAND RS FF.

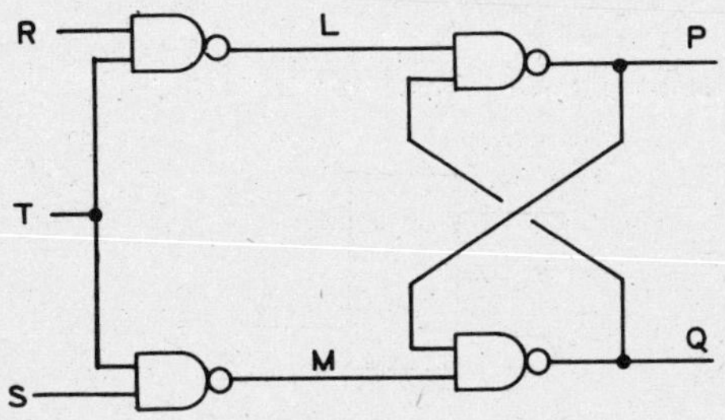


If T is at 0, then both L and M are at 1, and the outputs are inhibited from changing. If T is at 1, then L and M will be the inverse of R and S, and these will affect the outputs P and Q of the FF. When T reverts to a 0, L and M again become 1s. P and Q will become locked to their new conditions. A truth table will clarify this.

Assume the first line is the starting conditions.

| T | R | S | L | M | P | Q |
|---|---|---|---|---|---|---|
| 1 | 1 | 0 | 0 | 1 | 1 | 0 |
| 0 | 1 | 0 | 1 | 1 | 1 | 0 |
| 0 | 0 | 1 | 1 | 1 | 1 | 0 |
| 1 | 0 | 1 | 1 | 0 | 0 | 1 |
| 0 | 0 | 1 | 1 | 1 | 0 | 1 |
| 0 | 1 | 0 | 1 | 1 | 0 | 1 |

The student should follow each of these lines carefully.

Note that although the inputs R and S change simultaneously, L and M remain at 1 if T is a 0. Note that P and Q do not change until T becomes 1. Then these outputs are remembered when T reverts to 0.

The following illustrates the importance of this circuit. Suppose R and S do not change simultaneously. Suppose the disallowed input conditions of 1 and 1, or 0 and 0, momentarily occur. It does not matter if they do occur because they cannot affect the output until T becomes a 1. T is kept at 0 until the R and S inputs have settled down into opposite conditions. This whole circuit is called a *clocked RS FF*.

**Q:** What is a clocked RS FF?

**A:** It is one whose outputs are protected against disallowed input conditions. Its outputs can only change when a clock pulse is at the correct logic level at the T input.

A simple symbol for the clocked RS is this.

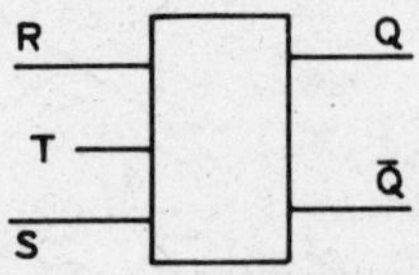


Its main properties are:

1. Both its outputs will always be in opposite states.
2. It cannot assume undesired output conditions as the basic RS can.
3. It can act as a true memory, as its outputs do not change at the same time as the inputs.

We will now examine three modifications to the clocked RS FF. Let us start with a 0 at A and a 1 at B.

**Q:** Draw a one-row truth table for this.

**A:** This is it.

| A | B | P | Q |
|---|---|---|---|
| 0 | 1 | 1 | 0 |

If the inputs now change simultaneously, then the 0 input at B will force a 1 to appear at Q. Gate A will now have both inputs at 1.

**Q:** From the truth table, what will its output be?

**A:** This is the only condition where the output can be 0.

So now we can draw a one-line truth table for this condition.

**Q:** Draw it.

**A:**

| A | B | P | Q |
|---|---|---|---|
| 1 | 0 | 0 | 1 |

A complete table for this FF can now be drawn.

| A | B | P | Q |
|---|---|---|---|
| 0 | 1 | 1 | 0 |
| 1 | 0 | 0 | 1 |

We have previously seen that two 1 inputs will cause no change, no matter what state the FF is in. Note that as we did not change input B, and P remained a 1, then nothing happened to alter output Q.

Let us summarize these changes at the input A with a truth table.

| A | B | P | Q |
|---|---|---|---|
| 0 | 1 | 1 | 0 |
| 1 | 1 | 1 | 0 |
| 0 | 1 | 1 | 0 |

Obviously, if we started with the inputs reversed so that A was a 1 and B a 0, and then changed input B, the results would be similar.

**Q:** Draw a truth table for this set of conditions.

**A:**

| A | B | P | Q |
|---|---|---|---|
| 1 | 0 | 0 | 1 |
| 1 | 1 | 0 | 1 |
| 1 | 0 | 0 | 1 |

Note that columns P and Q are reversed from the above. Suppose now that both inputs A and B are set at 0, what are the outputs now?

**Q:** Look at the truth table for the NAND and determine the outputs.

**A:** They are both 1s.

For various reasons, this condition is avoided in digital electronics. Having both outputs of a FF the same is not usually desired.

The interesting action in this FF circuit occurs when both inputs are the opposite of each other, and when they both change simultaneously.

The first modification is to include an inverter, as in this figure, and to supply only one input.

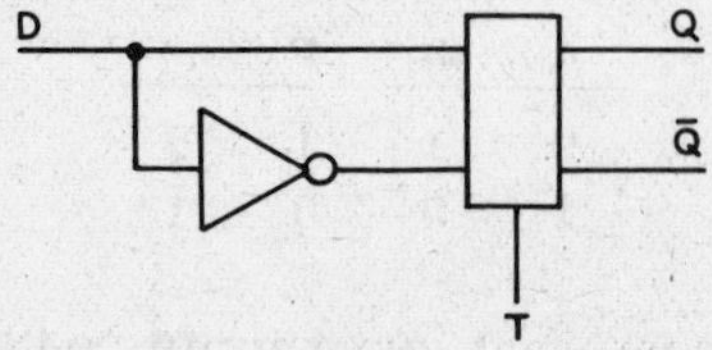


Whatever is presented as an input at point D will pass to input R and be inverted at input S.

**Q:** Does this ensure that opposite conditions are always present at the input of the FF?

**A:** Yes.

This input will then affect the output when the clock pulse arrives at T. The input can change continually, but it will not affect the output until the clock pulse at T goes high.

**Q:** When T is a 1, what happens to the output if the input is changing?

**A:** It will also change.

**Q:** When T goes low, what will the output become?

**A:** It will remember the input.

In fact, the output will be a memory of the input conditions which existed at the time the T input went to 0.

So this circuit remembers its last input. This is called a *"D" FF*. It is used a great deal as it is a very simple memory circuit.

**Q:** What is a common use for the "D" FF?

**A:** It is a very simple memory circuit.

The data to be remembered is entered at the D input. D stands for data, hence the name of the FF. When the clock pulse arrives at T, the data is transferred to the Q output. A truth table for the type D FF is this.

| T | D | Q |
|---|---|---|
| 1 | 1 | 1 |
| 0 | 1 | 1 |
| 0 | 0 | 1 |
| 1 | 0 | 0 |
| 0 | 0 | 0 |
| 0 | 1 | 0 |
| 1 | 1 | 1 |

The output follows the D input only when the T goes high. A second modification to the clocked RS is shown here.

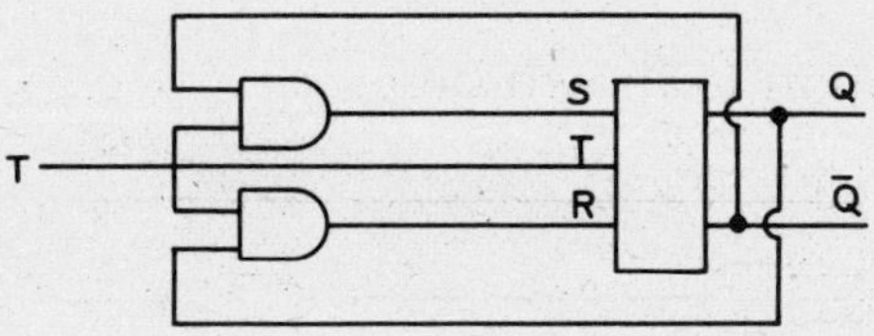


Note that this has only one input, the T. There are not separate data inputs. Its operation is very simple. Every time the T input goes to a 1, the output changes state. Rather than use a truth table, this is best shown by using a voltage time graph, or timing diagram.

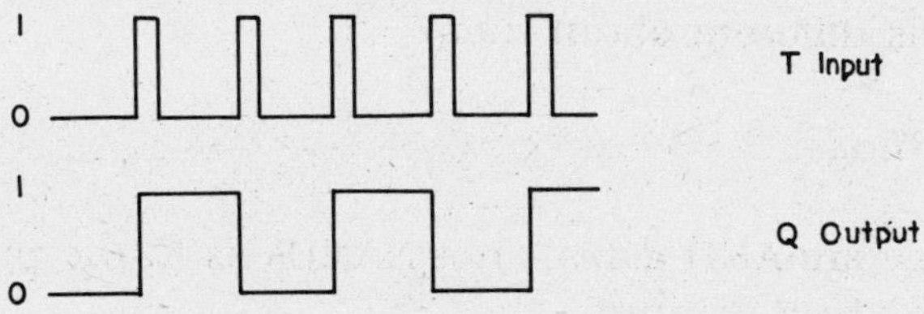


This becomes very interesting and important when the T input is a square wave (see diagram on next page).

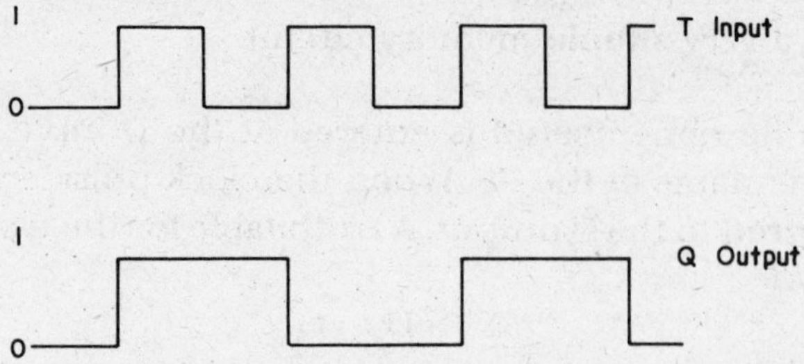


The Q output has only half as many changes as the T input. In other words, it is half the input frequency.

A very important way of saying this is: The input has been divided by 2. This is very important in computers and digital counting. We will return to this later. This type of FF is called a *"T" flip-flop*.

**Q:** Why is the "T" flip-flop important?

**A:** Because it divides the input by 2, and can thus be used in binary counting.

The third modification is shown here.

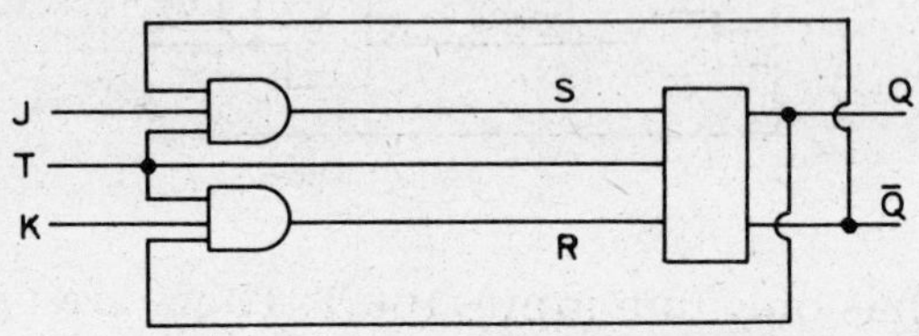


Note that here the new gates are slightly different.

**Q:** What is different about them?

**A:** Two things:

1. They are AND gates—not NANDs as in the previous FFs
2. They have 3 inputs

They are called 3-input AND gates. The two data inputs are labelled J and K. That is why this circuit is called the *"J-K" FF*. It is the most versatile of all FFs.

The truth table for the JK FF is this.

| J | K | $Q_{n+1}$ |
|---|---|---|
| 0 | 0 | Inhibit |
| 0 | 1 | 0 |
| 1 | 0 | 1 |
| 1 | 1 | Toggle |

This table needs a little explanation as it is different from the others seen before. The column headings are defined as follows:

1. J and K mean the inputs to the J and K inputs. These will be either 0 or 1.
2. $Q_{n+1}$ means the Q output AFTER the clock pulse has returned to 0.

This FF will stay in a given state until a clock pulse is supplied to the T input. Any changes occurring at the J and K inputs will NOT affect the outputs until the clock input goes to a 1.

When the clock input is at 1, any changes on the inputs will be immediately reflected at the outputs. Hence, the inputs are kept steady when the clock input is at 1. When the clock input goes to 0, then the FF is inhibited; this means the outputs are not affected by changes at the inputs.

**Q:** What is the condition which must be met for the outputs of the JK FF to change?

**A:** The clock pulse must be at a 1 level.

**Q:** What happens to the outputs when the clock is at 0?

**A:** They are prevented from changing, that is, the FF is inhibited.

To return to the truth table:

**Line 1.** This means that both the J and the K inputs are at 0. And after the clock pulse has arrived and gone (that is, the T input has gone from 0 to 1 and then back to 0 again), the outputs were inhibited from changing. In other words, the outputs remained whatever they were before the clock pulse.

**Line 2.** If J = 0 and K = 1, then the Q output will become 0 when the clock pulse arrives. If it was 0 before the clock pulse, it will remain a 1.

**Line 3.** If J = 1 and K = 0, then the Q output will become 1 when the clock pulse arrives. If it was 1 before the clock pulse, it will remain a 1.

**Line 4.** If both J and K are 1, then the output will change state when the clock pulse arrives; ie., it will toggle.

A comment for the $\overline{Q}$ output is not needed as it is always the inverse of the Q output.

**Q:** Cover up the column on the right-hand side of the answer (labeled "Q"). The two left-hand columns are a succession of inputs at the J and K inputs. The corresponding entry in the covered right-hand column is the appropriate output *after* the clock pulse.

Write out what you think is the correct entry for the covered column, check it, and then try the next line.

**A:**

| J | K | Q |
|---|---|---|
| 1 | 0 | 1 |
| 1 | 1 | 0 |
| 1 | 0 | 1 |
| 0 | 1 | 0 |
| 1 | 1 | 1 |
| 1 | 0 | 1 |
| 0 | 1 | 0 |
| 0 | 0 | 0 |
| 1 | 1 | 1 |
| 0 | 0 | 1 |
| 1 | 0 | 1 |
| 0 | 1 | 0 |
| 1 | 0 | 1 |
| 1 | 1 | 0 |

The next two questions are really exercises. The information has not been stated explicitly in the text, but it can be deduced from the truth table for the JK FF. The thinking and reasoning required here are typical of that needed in digital design and analysis of circuits.

**Q:** Can you arrange the inputs of a JK FF so that the outputs will behave like a D FF? (Hint: an extra external gate or inverter is needed.)

**A:** The circuit is this.

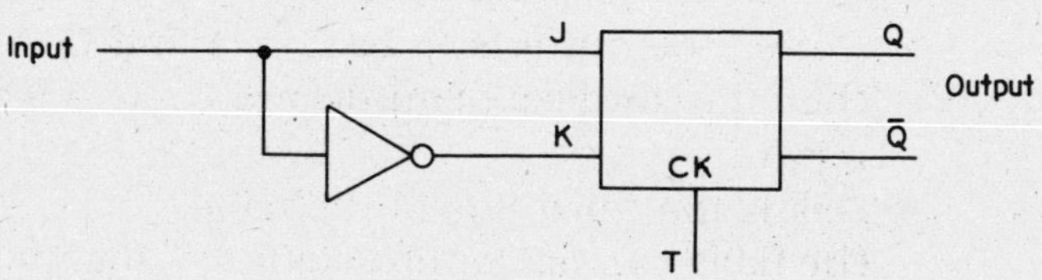


**Q:** Work out the truth table for this and verify that it is behaving like a D FF.

**A:** This is the correct truth table.

| Input | J | K | $Q_{n+1}$ |
|---|---|---|---|
| 1 | 1 | 0 | 1 |
| 0 | 0 | 1 | 0 |

No other conditions can occur.

**Q:** Can you arrange the inputs and output connections of a JK FF so that it will behave like a T FF?

**A:** There are two ways of doing this; they are both shown here.

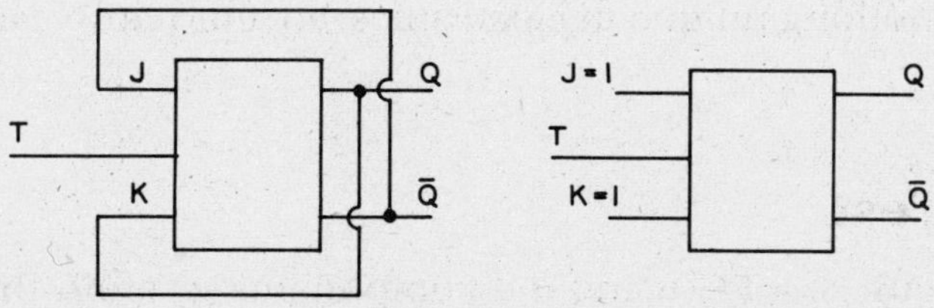


**Q:** Verify this is working like a T FF by working out the truth table.

**A:** The table for the left-hand diagram can be worked out in this manner:

Assume the initial outputs condition are Q = 1.

Then the initial inputs are $J = \overline{Q} = 0$ and $K = Q = 1$.
This is the first line in the table.
After the first clock pulse, these inputs will force the outputs to be $Q = 0$ and $\overline{Q} = 1$.
Then the second inputs are $J = \overline{Q} = 1$ and $K = Q = 0$.
This is the second clock pulse; these inputs will force the outputs to be $Q = 1$ and $\overline{Q} = 0$.
Then the third set of inputs are $J = \overline{Q} = 0$ and $K = Q = 1$.
This is the third line of the table.
The table can be continued in this manner until the pattern is definitely established.
This is the table.

| J | K | $Q_{n+1}$ |
|---|---|---|
| 0 | 1 | 0 |
| 1 | 0 | 1 |
| 0 | 1 | 0 |

Remember that the $Q_{n+1}$ column is the Q output AFTER the clock pulse and it results from the input conditions printed in the same line.

The right-hand drawing is much easier to work out. Look at the truth table for the simple JK FF. With both J and K at 1, it will toggle. This is what a T FF does.

The student should learn two things from the above exercises. First, a step-by-step approach to working out problems will always be successful, no matter how difficult the problem may appear at first. Second, there is nothing unique in electronics; an alternative can always be found.

## Self-Test

Draw the circuit of a FF, using the components given in the following questions.

1. Two transistors
2. Two inverters, using the logic symbols. Draw the truth table.
3. Two NAND gates to make an RS FF. Do the truth table.
4. Two NOR gates to make an RS FF. Do the truth table.
5. A clocked RS, using gates. Do the truth table.

6. A T FF with truth table
7. A D FF with truth table
8. A JK FF with truth table
9. What is a T FF used for?
10. Why is the JK FF valuable?

### Answers

All the above answers can be found in the text.

# Digital Waveforms and Pulses

The voltage levels in digital electronics are usually represented by the symbols 1 and 0. The 1 usually indicates a voltage, above ground, and the 0 is used for ground or 0V.

In digital electronics typical graphs of voltage variations with time look like the patterns shown in this figure.

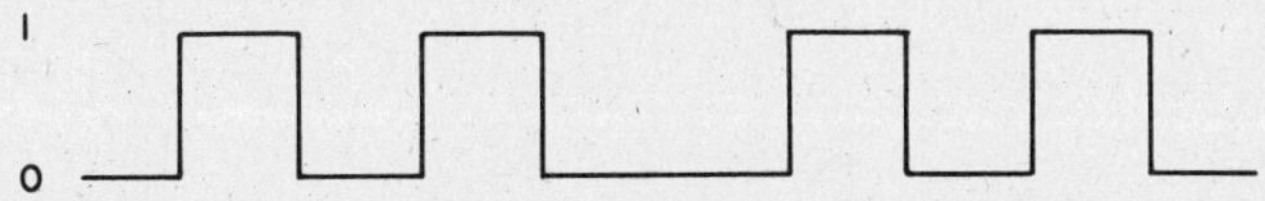


In their simplest form, these patterns can be generated by the opening and closing of a mechanical switch; but often they are generated by digital circuits. A single voltage jump is called a *pulse*, and many are called a *pulse train*.

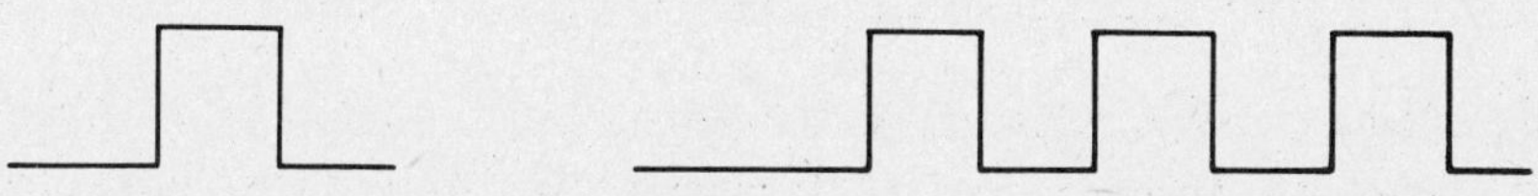

These voltage patterns are often called *waveforms*, and they are very common in digital electronics. They will be introduced in this section along with these other simple ideas:

1. The simple use of the oscilloscope in viewing voltages and pulses
2. The concept of the pulse
3. Some definitions of the important points about pulses
4. Some definitions about pulse trains.

To fully appreciate these voltage patterns or waveforms, they have to be seen—and the only way to do this is to use the oscilloscope.

The oscilloscope is *the* instrument of electronics. This section, more than any other, requires that the student use the "scope" in viewing the waveforms.

Due to space requirements, scope operation will not be covered. It is better that the student be shown how to use a scope by an instructor or an experienced user. A few simple experiments with the scope are now described. These serve the double purpose of introducing the scope and introducing the concept of viewing pulses and waveforms.

### *Experiment 1: How to measure battery voltage with the scope*

Procedure:

1. Adjust the INTENSITY control and the POSITION control so that the trace is centered on the screen.
2. Set the HORIZONTAL TIME to 1 cm/sec. The dot should now travel slowly from left to right.
3. Set the VERTICAL VOLTAGE control to 5 v/cm.
4. Put the probe ground clip on the negative terminal of the battery.
5. Put the probe "tip" onto the positive terminal of the battery.
6. The trace should jump like this.

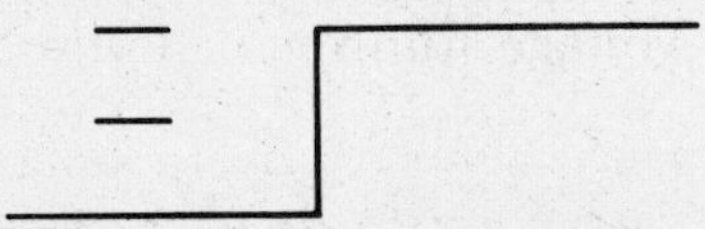

   Check the voltage measurements.

7. Remove the probe and observe the trace drop back to the center of the screen.
8. Put the probe on and off the battery several times. Observe the trace do this.

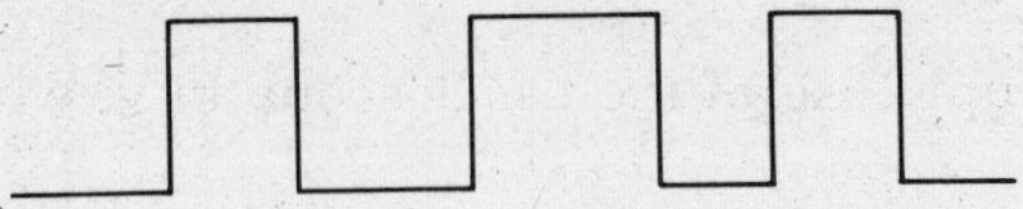

A voltage jump such as seen in this experiment is called a *pulse*. It is a POSITIVE pulse. Putting the probe on and off the battery produced a rough *pulse train*.

### *Experiment 2*

Repeat Experiment 1 with the battery reversed. This will produce NEGATIVE pulses, like this.

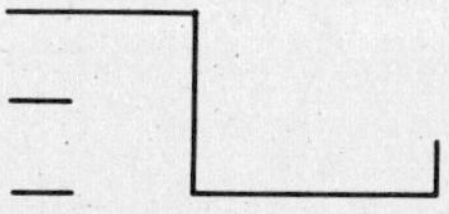

At this time a few definitions about pulses are needed. Every pulse has a "leading edge" and a "trailing edge". These are shown for positive and negative pulses in this figure.

The time between the two edges is referred to as the "pulse duration".

**Q:** What is the pulse duration—or the PD?

**A:** The PD is the time a pulse lasts; in other words, the time between its leading and trailing edges.

In both the above experiments the pulse duration was dependent upon the length of time the probe was on the battery terminals. In a similar manner, the duration of a pulse could be dependent upon the manual operation of a mechanical switch.

Obviously, very fast or accurately timed pulses cannot be produced in this manner. If an accurate train of pulses is required whose duration and repetition must be exact, then an electronic instrument known as a *pulse generator* must be used.

**Q:** What is a pulse generator?

**A:** An electronic instrument which produces accurate pulse trains.

A simple pulse generator is found in most scopes. It is called the *calibrate* output. It is used to calibrate the probes for correct measurements. Usually it puts out a train of pulses whose voltage level can be varied.

The next experiment will use this output.

***Experiment 3: To observe and measure a train of pulses***

Procedure:

1. Set the HORIZONTAL time control to 1 ms/cm. The dot will now look like a continuous line.
2. Center this line on the screen.
3. Put the probe into the CALIB OUT socket.
4. View on the screen this pattern.

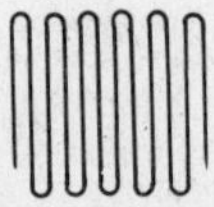

5. Adjust the TRIGGER control to get this pattern. (You may need help with this adjustment.)

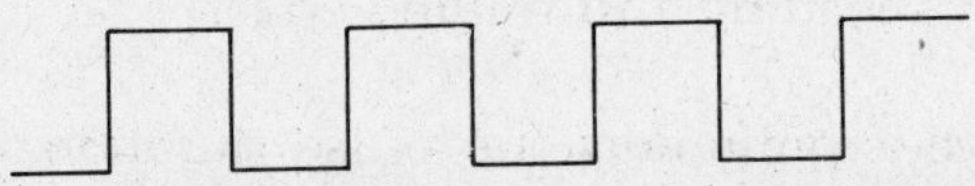

6. Measure the voltage levels and the times, as shown here. Your voltage level can be altered by the CALIB control on the scope.

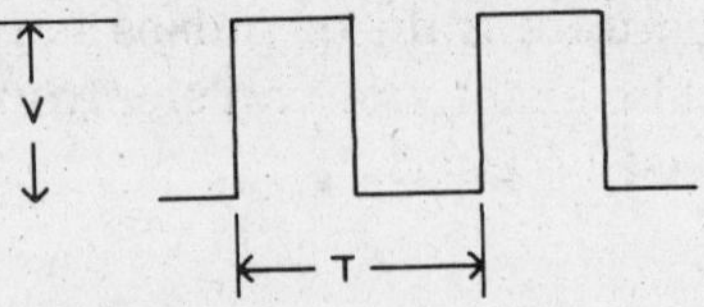


What you are viewing is a "squarewave" pattern or pulse train.

The important measurements in a pulse train are shown in the following figure.

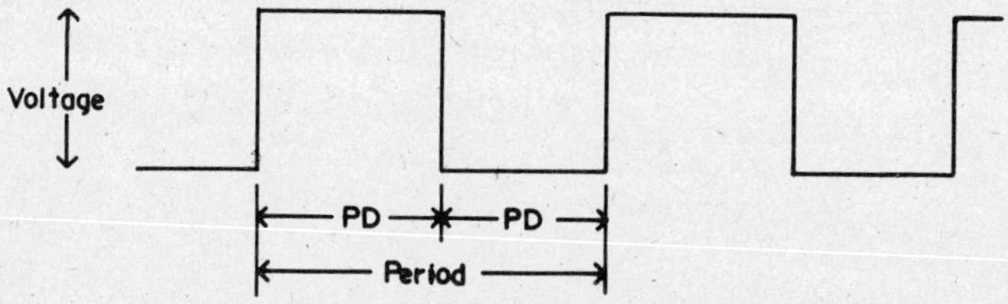


As the pulses are continuously produced, then a certain number will be produced in a given time. The usual way of assessing this is to decide how many pulses are produced in a second. The number produced in a second is called the *frequency* of the pulses. For example, if 10 pulses are produced in 1 second, then the frequency is 10 pulses per second. Pulses per second is usually called "*Hertz*". So the frequency would be 10 Hertz, or 10 Hz.

**Q:** If 100 pulses are produced in one second, what is the frequency?

**A:** 100 Hz.

**Q:** Suppose we have 1,000 pulses per second. What now is the frequency?

**A:** 1,000 Hz. Also called 1 kHz.

**Q:** What is the frequency of a million pulses per second?

**A:** 1,000,000 Hz, also called 1 Megahertz or 1 MHz.

The frequency is also called the *pulse repetition rate*. This term is sometimes preferred in digital work, leaving the term frequency for AC work.

**Q:** What does pulse repetition rate mean?

**A:** It is another term for frequency.

The term *period* of a pulse is often used when describing a pulse train. The following figure shows this.

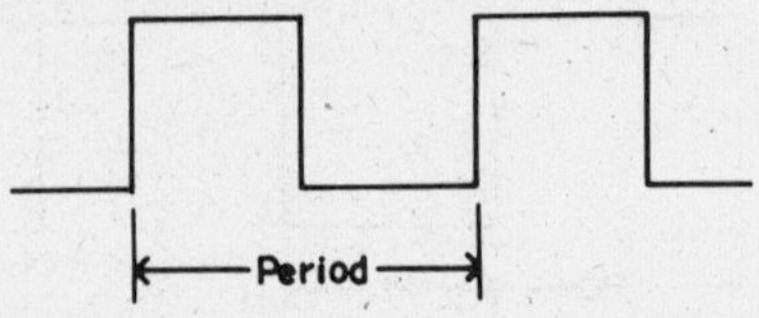


A simple mathematical relation exists between period and frequency:

Period = 1/frequency

Example: What is the period of a 1 kHz pulse?

Period = 1/1000 seconds
= 1 millisecond
= 1 ms

Example: What is the frequency of a pulse train with a period of 10 ms?

Frequency = 1/Period
= 1/10 ms
= 100 Hz

This later example is typical of what is required when using a scope. The period of the waveform can be read from the screen, and then the frequency must be found from that.

The student should do the few following examples to gain familiarity with this.

The answers to the following questions are in the right-hand column. Cover them up as you work on the questions.

**Q:** Find the period of the pulses which occur at these frequencies:

| | *Answers* |
|---|---|
| 1. 10 Hz | 0.1 secs |
| 2. 100 Hz | 0.01 secs |
| 3. 330 Hz | 3 ms |
| 4. 2 kHz | 0.5 ms |
| 5. 5 kHz | 0.2 ms |
| 6. 10 kHz | 0.1 ms |
| 7. 12 kHz | 0.083 ms |
| 8. 50 kHz | 0.02 ms |
| 9. 100 kHz | 0.01 ms |
| 10. 1 MHz | 0.001 ms or 1 us |

**Q:** Find the frequency of the pulses which have these periods:

| | *Answers* |
|---|---|
| 11. 0.1 secs | 10 Hz |
| 12. 1 secs | 1 Hz |
| 13. 10 secs | 0.1 Hz |
| 14. 1 ms | 1 kHz |
| 15. 0.1 ms | 10 kHz |
| 16. 0.5 ms | 2 kHz |
| 17. 0.33 ms | 3 kHz |
| 18. 1 microsecs | 1 MHz |
| 19. 0.3 microsecs | 3.3 MHz |
| 20. 0.001 microsecs | 1,000 MHz |

CHAPTER

7

# Advanced Digital Circuits

**OBJECTIVES:** Upon completion of Chapter 7 you will be able to do the following:

- ☐ Explain the main advantage of binary over decimal
- ☐ Work experiments utilizing gates for addition and subtraction
- ☐ Do experiments in Binary-to-Decimal conversion
- ☐ Do experiments in Decimal-to-Binary conversion
- ☐ Explain the difference between volatile and nonvolatile memories
- ☐ Explain the two operations performed on a RAM
- ☐ Explain the difference between an EPROM and an EEPROM
- ☐ Explain the type of memory that holds data and cannot be written into, which is used for very large-volume applications
- ☐ Explain what type of circuit converts one DC voltage level to another DC voltage level

EVERYTHING IN ELECTRONICS is useful. Something which is not is quickly discarded for an alternative that is. This is because electronics is a technology; in other words, it is a branch of a science which is used for some practical purpose.

All the circuits of the previous chapters can be put to good use, and will perform many desirable tasks. This chapter presents the simplest of these.

In this chapter, you will see:

1. How gates will add and subtract
2. How FFs will count
3. What Electronic Semiconductor Memories are
4. What Serial-to-Parallel conversion is
5. What Parallel-to-Serial conversion is
6. How to do Binary-to-Decimal conversion
7. How to do Decimal-to-Binary conversion

The simple operations of addition and subtraction are very important in computers and automatic equipment. In the decimal system, they can easily be achieved by mechanical wheels with numbers on them. The car odometer, cash register, and older types of calculating machines are good examples.

In modern electronic equipment the binary number system is used. This is because it is easier to implement binary numbers with electronic equipment, and electronic binary operations can be made to occur much faster than decimal operations with mechanical wheels.

**Q:** What is the main advantage of binary over decimal?

**A:** It is easier to implement electronically and it is very much faster.

If we look at binary addition, we will soon find out how easy it is to effect with simple gates. First, consider the rules of binary addition.

| | | | | |
|---|---|---|---|---|
| 0 | 0 | 1 | 1 | row A |
| +0 | +1 | +0 | +1 | row B |
| 0 | 1 | 1 | 10 | row Y |

**Q:** What is different about the last column?

**A:** It has a carry.

Suppose we rearrange these additions in the form of a truth table.

| A | B | Y | |
|---|---|---|---|
| 0 | 0 | 0 | 1st column |
| 0 | 1 | 1 | 2nd column |
| 1 | 0 | 1 | 3rd column |
| 1 | 1 | 0 | 4th column (ignoring the carry) |

**Q:** Look at the truth table. What binary function is it?

**A:** The Exclusive OR.

This means an Exclusive OR circuit will perform simple binary addition.

**Q:** If we use an Exclusive OR for binary addition, what is missing?

**A:** The carry needed in the fourth case.

This carry can easily be supplied. Look at this circuit.

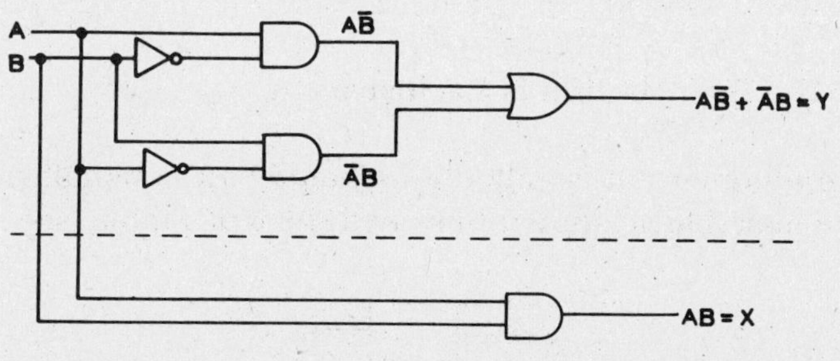


The part above the dotted line is an Exclusive OR circuit. The part below is a simple AND gate. The SUM output is given by Y, and the carry output is given by X, which is simply the ANDing of the two inputs.

**Q:** As an exercise, try to work out the sum part of the above circuit using NAND gates only. First, use algebra to produce an expression for S which contains only ANDs. Invert this to give an expression for S which has a NOT bar over it.

**A:** $S = A\bar{B} + \bar{A}B$, so $\bar{S} = \overline{A\bar{B} + \bar{A}B}$.
Use De Morgan's Laws to give:

$$\bar{S} = \overline{A\bar{B}} \cdot \overline{\bar{A}B}$$

Invert to get:

$$S = \overline{\overline{A\bar{B}} \cdot \overline{\bar{A}B}}$$

Use this to make a circuit.

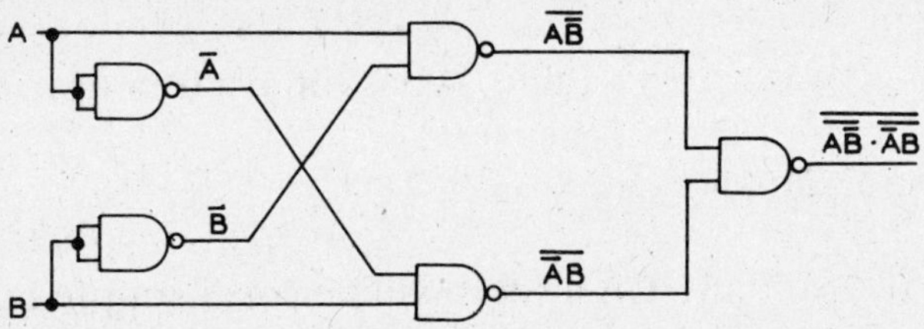


**Q:** Draw a truth table for this expression and check that it is an Exclusive OR.

**A:**

| A | B | $\bar{A}$ | $\bar{B}$ | $\bar{A}B$ | $A\bar{B}$ | $\overline{\bar{A}B}$ | $\overline{A\bar{B}}$ | $\overline{\bar{A}B} \cdot \overline{A\bar{B}}$ | $\overline{\overline{\bar{A}B} \cdot \overline{A\bar{B}}}$ |
|---|---|---|---|---|---|---|---|---|---|
| 0 | 0 | 1 | 1 | 0 | 0 | 1 | 1 | 1 | 0 |
| 0 | 1 | 1 | 0 | 1 | 0 | 0 | 1 | 0 | 1 |
| 1 | 0 | 0 | 1 | 0 | 1 | 1 | 0 | 0 | 1 |
| 1 | 1 | 0 | 0 | 0 | 0 | 1 | 1 | 1 | 0 |

The final column is Exclusive OR.

The complete circuit is called a *half adder*. It is usually represented by a single logic block, shown here with its truth table (see next page).

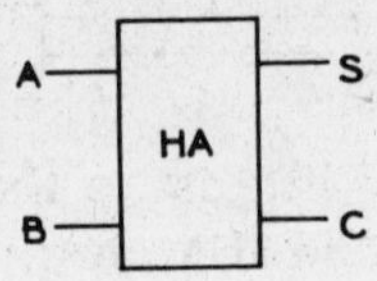

| A | B | S | C |
|---|---|---|---|
| 0 | 0 | 0 | 0 |
| 0 | 1 | 1 | 0 |
| 1 | 0 | 1 | 0 |
| 1 | 1 | 0 | 1 |

Because of their widespread use in computers, they are manufactured as complete units and do not need to be built from gates by the user or designer.

The half adder has only two inputs which it can add together. If three inputs are required then this circuit can be used. This is called a *full adder*.

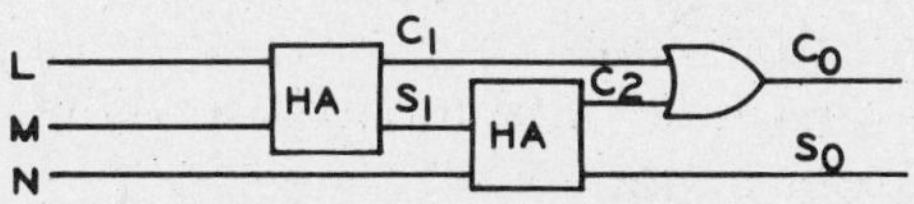


This is the truth table for the full adder.

| L | M | N | $C_1$ | $S_1$ | $C_2$ | $C_0$ | $S_0$ |
|---|---|---|---|---|---|---|---|
| 0 | 0 | 0 | 0 | 0 | 0 | 0 | 0 |
| 0 | 0 | 1 | 0 | 0 | 0 | 0 | 1 |
| 0 | 1 | 0 | 0 | 1 | 0 | 0 | 1 |
| 0 | 1 | 1 | 0 | 1 | 1 | 1 | 0 |
| 1 | 0 | 0 | 0 | 1 | 0 | 0 | 1 |
| 1 | 0 | 1 | 0 | 1 | 1 | 1 | 0 |
| 1 | 1 | 0 | 1 | 0 | 0 | 1 | 0 |
| 1 | 1 | 1 | 1 | 0 | 0 | 1 | 1 |

By connecting full and half adders together the addition of large numbers is possible; however, these circuits will not be covered here. Binary subtraction can be approached in the same manner.

| | | | | |
|---|---|---|---|---|
| 0 | 1 | 0 | 1 | row A |
| −0 | −0 | −1 | −1 | row B |
| 0 | 1 | −1 | 0 | row X |

**Q:** Write this out in truth table form.

**A:**

| row A | row B | row X |
|---|---|---|
| 0 | 0 | 0 |
| 1 | 0 | 1 |
| 0 | 1 | 1 |
| 1 | 1 | 0 |

**Q:** What function is this?

**A:** Again, it is Exclusive OR.

So it would appear that subtraction can be affected by the same circuit as addition, provided the negative sign can be recognized. We will not cover this further here.

From the chapter on binary arithmetic you will remember that other methods of binary subtraction exist.

**Q:** What are these other methods?

**A:** 1s and 2s complement methods.

There are some complications using these, as we have to identify and use end-carry-arounds, and perform inversions.

**Q:** Are inversions easy or hard to effectuate or realize in digital electronics?

**A:** Very easy; just use an inverter.

Processing end-carry-arounds, etc. are a little harder and we will not cover those here. In summary, it should be obvious that subtraction can be performed easily.

Counting is a very important function in binary arithmetic and is used extensively in computers and control equipment. Binary counting is performed with FFs. To make the explanations easy, we will use the T FF only.

**Q:** Draw the circuit symbol for a T FF.

**A:**

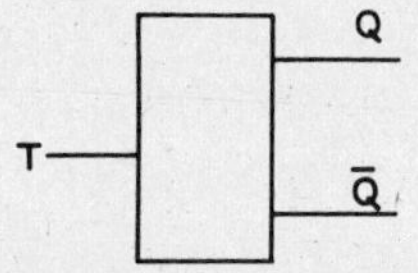


**Q:** What conditions can the output take, and how is it made to change?

**A:** The outputs are either 1 or 0. They change every time a clock pulse arrives at the T input.

A voltage-time graph or a timing diagram will illustrate this. For convenience, we will assume the output changes when the T input goes NEGATIVE. Many FFs work in this way.

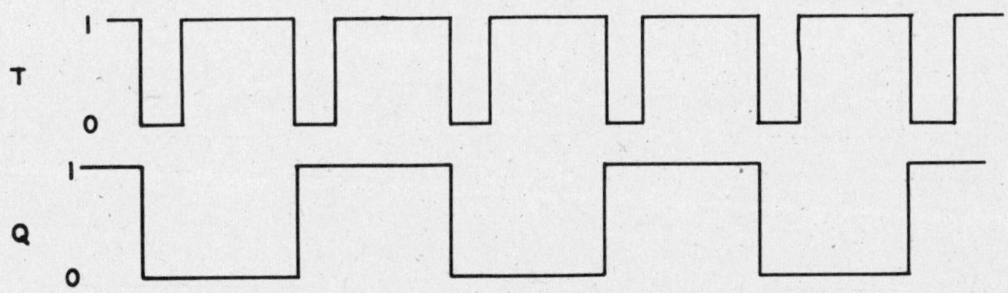


Notice that the Q output goes negative on the first, third, fifth . . . and all subsequent odd counts of the input pulses. So Q gets back to the starting state every second input pulse.

If we count the positive transistions of the Q output, there will be half as many as there are input pulses. So, in effect, the input has been divided by 2. Another way of looking at this is the following. Every time the Q output goes negative, there have been 2 input pulses. So the Q output is counting the inputs 2 at a time.

**Q:** What number system counts things in 2s?

**A:** Binary.

So a single FF used in this manner is a simple binary counter. Let us extend this idea by connecting another FF as shown.

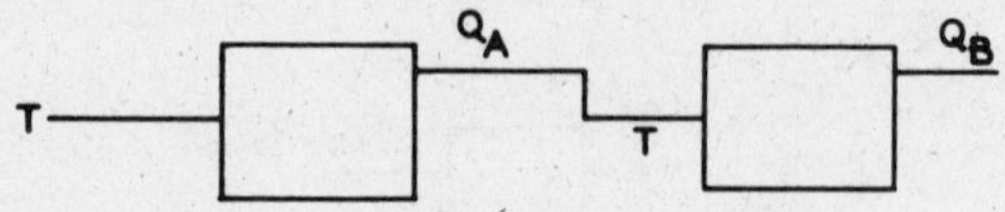


The output of the second FF, labelled $Q_B$, will change only when the input goes negative. Therefore, $Q_B$ will change state only when $Q_A$ goes negative.

**Q:** Draw a timing diagram of the input at T and the outputs at $Q_A$ and $Q_B$.

**A:** This is it.

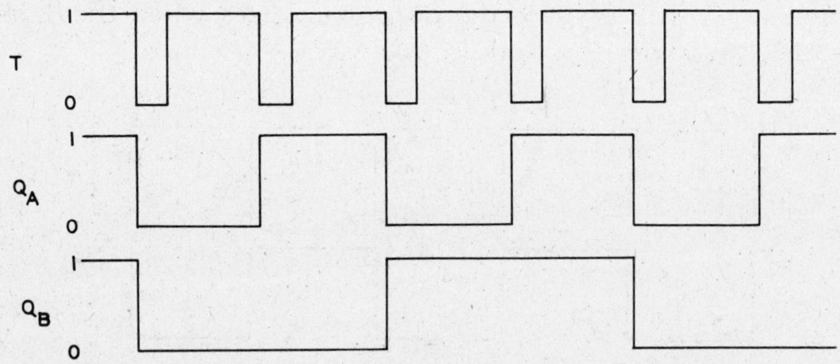


Look at these waveforms and try to answer this question.

**Q:** How many input pulses must occur before $Q_B$ reverts back to its starting condition?

**A:** 4.

So the $Q_B$ output effectively divides by 4. Now we will put three FFs in series and see what happens.

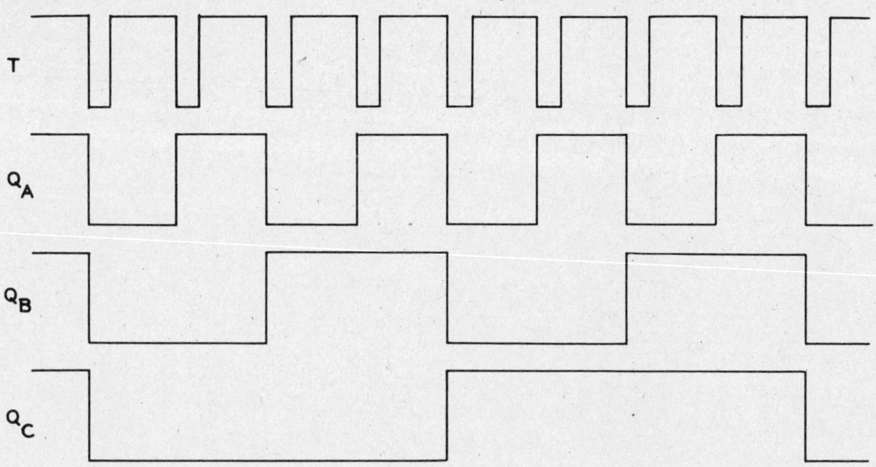


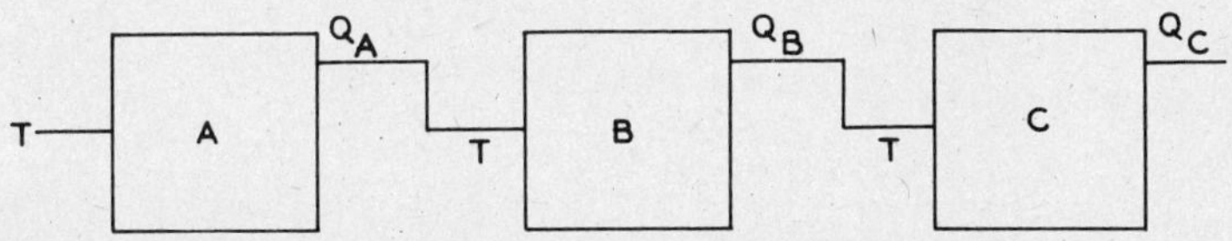


**Q:** Count the input pulses. What are we dividing by now?

**A:** 8.

For every FF added to the chain, the frequency of the input pulses is divided again by 2. The division ratio is given by $2^N$, where N is the number of FFs used in the chain.

**Q:** By what number are we dividing if 6 FFs are used?

**A:** Here $N = 6$. So $2^N = 2^6 = 64$.

**Q:** By what number are we dividing if 10 FFs are used?

**A:** Here $N = 10$. So $2^N = 2^{10} = 1{,}024$.

**Q:** We wish to divide the input by 512. How many FFs are needed?

**A:** Here $2^N = 512$. So $N = 9$.

**Q:** Write out the powers of $2^N$ for $N = 1$ up to $N = 12$.

**A:** When $N = 1$ then $2^N = 2^1 = 2$

| | |
|---|---|
| 2 | 4 |
| 3 | 8 |
| 4 | 16 |
| 5 | 32 |
| 6 | 64 |
| 7 | 128 |
| 8 | 256 |
| 9 | 512 |
| 10 | 1,024 |
| 11 | 2,048 |
| 12 | 4,096 |

With familiarity, these numbers become automatic.

Look at this chain of 3 FFs. It has been drawn from right to left instead of left to right. Thus, the input is applied to the right-hand FF. The reason for this will be apparent in a moment.

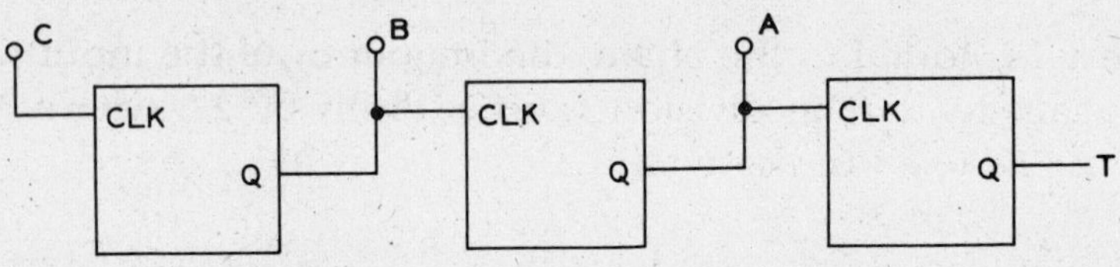


Below are shown the waveforms.

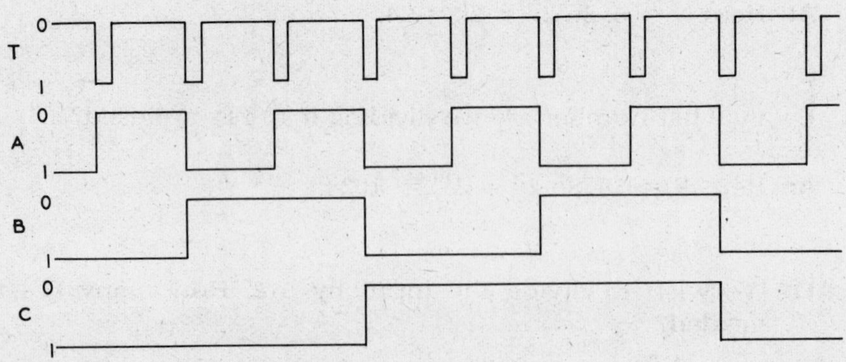

We will now examine these waveforms and determine the state of each output after each pulse. Start at the dotted line where all outputs are at 1.

**Q:** Write down the output states in a truth table. The columns are to be the outputs. Hint: Use columns headed C, B, and A.

**A:**

| C | B | A | |
|---|---|---|---|
| 1 | 1 | 1 | Start |
| 1 | 1 | 0 | After 1st pulse |
| 1 | 0 | 1 | 2nd |
| 1 | 0 | 0 | 3rd |
| 0 | 1 | 1 | 4th |
| 0 | 1 | 0 | 5th |
| 0 | 0 | 1 | 6th |
| 0 | 0 | 0 | 7th |

**Q:** What does this table display?

**A:** It is binary counting in reverse. It starts at 7 and counts down to 0.

Counting backwards in binary can be achieved by connecting the Q to each successive T input of the FF chain. In each of the FF chains used in the past frames, the input pulses were applied to the first FF only. The others in the chain were triggered only when the output of the previous FF changed in a direction which would cause triggering. All the FFs are independently triggered, and they all operate at different times.

This type of FF operation is known as *asynchronous* operation. These counter chains are called *asynchronous counters*.

**Q:** What is an asynchronous counter?

**A:** It is a chain of FFs used for counting, where the pulses to be counted are applied to the first FF in the chain only. The other FFs are triggered successively in a binary count from the outputs of the previous FF in the chain.

We will now look at a very important use for FFs. Consider this chain of JK FFs.

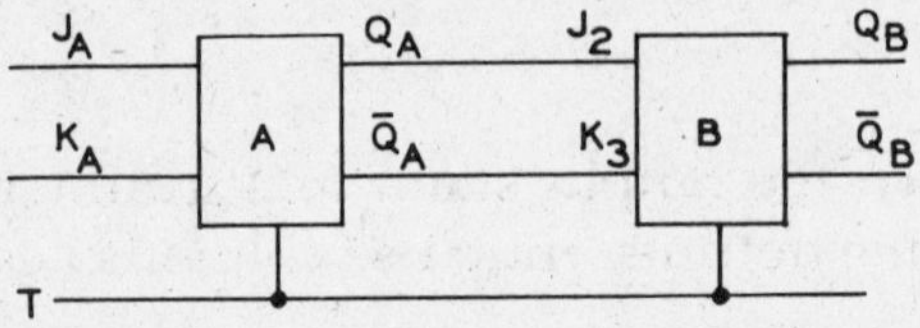


Note that the T inputs or clock inputs are all connected together, so that the clock pulses are applied to all the FFs simultaneously. This is known as *synchronous* operation.

**Q:** What is synchronous operation of a chain of FFs?

**A:** It is where the clock or toggle pulses are applied to all the FFs in a chain simultaneously.

**Q:** Draw the truth table for a JK FF. We will need to refer to it in the following discussion.

**A:**

| J | K | Q |
|---|---|---|
| 0 | 0 | Inhibit |
| 0 | 1 | 0 |
| 1 | 0 | 1 |
| 1 | 1 | Toggle |

We will start with $J_A = K_A = 0$. This will inhibit all the FFs. Also, we will have all the Q outputs at 0. Let $J_A$ go to a 1. Now the first T pulse arrives.

**Q:** What happens to $Q_A$? Look at the truth table.

**A:** It goes to a 1.

**Q:** What happens to all the other Qs in the chain?

**A:** They all stay at 0, as their J inputs are 0.

After the first T pulse has gone, let $J_A$ revert to 0. Now the second T pulse arrives.

**Q:** What happens to $Q_A$?

**A:** It goes back to 0, as the J input is 0.

But when the second pulse arrived, the J input to the B FF was a 1. This 1 is passed onto the Q output of the B FF. Now $J_A$ is kept at 0 and the third pulse arrives. $Q_B$ will go to 0, as its J input is 0. $Q_C$ will go to 1, as its J input is 1; it was the 1 at $Q_B$ before the pulse arrived.

What has happened with each successive pulse? The 1 which was present at the beginning of the sequence has been passed in turn from the $Q_A$ output to the $Q_B$ and then to the $Q_C$. Further pulses will pass it onto any other FFs in the chain. This can be shown graphically.

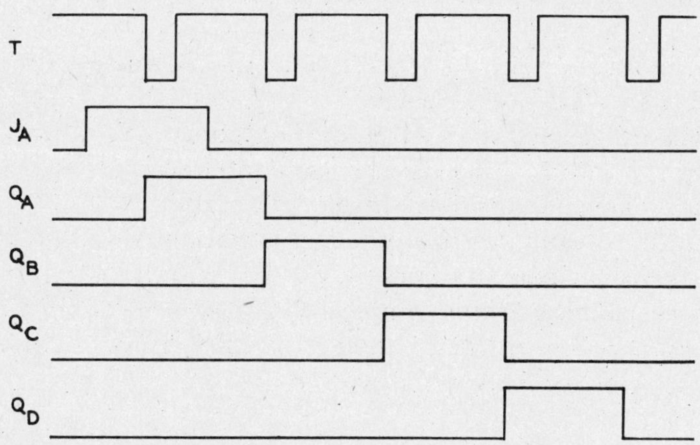


**Q:** Try to draw the outputs in truth table form.

**A:**

| A | B | C | D | |
|---|---|---|---|---|
| 0 | 0 | 0 | 0 | Start |
| 1 | 0 | 0 | 0 | After 1st pulse |
| 0 | 1 | 0 | 0 | 2nd |
| 0 | 0 | 1 | 0 | 3rd |
| 0 | 0 | 0 | 1 | 4th |
| | etc. | | | |

If at some later time $J_A$ was allowed to go to 1 again, this would also be entered and would pass down the line of FFs. The following table shows this.

| | A | B | C | D | E | F . . . | |
|---|---|---|---|---|---|---|---|
| | 0 | 0 | 0 | 0 | 0 | 0 | |
| 1st 1 at $J_A$ | 1 | 0 | 0 | 0 | 0 | 0 | |
| | 0 | 1 | 0 | 0 | 0 | 0 | 1st 1 passing down line of FFs |
| 2nd 1 at $J_A$ | 0 | 0 | 1 | 0 | 0 | 0 | |
| | 1 | 0 | 0 | 1 | 0 | 0 | |
| | 0 | 1 | 0 | 0 | 1 | 0 | |
| | 0 | 0 | 1 | 0 | 0 | 1 | |
| 2nd 1 passing down line of FFs | 0 | 0 | 0 | 1 | 0 | 0 | |

etc.

If F was the last FF in the line, then its Q output would revert to 0 as the first 1 was passed, and it would go to a 1 as the second 1 was passed.

In this way any pattern of 1s and 0s at the $J_A$ input can be passed along until it appears at the last FF.

If the clock pulses are stopped then the pattern will remain static on the FF outputs. In other words, it is remembered by them. A chain of FFs operated like this is thus a simple "memory".

If all the FF outputs are examined simultaneously, then the whole pattern of 1s and 0s can be read and used at one time. When used like this, they are called *parallel* outputs. The pattern came into the FF chain one bit at a time. This is called in *serial* form. A chain of FFs can also act as a *serial-to-parallel* converter.

**Q:** We have just described two simple uses for a chain of flip-flops. What are they?

**A:** Serial to parallel conversion, and memory.

Both of these applications are very important in computers and automatic control equipment. There are many variations of these two circuit applications, but what is shown here is the fundamental ideas on which they are built.

A pattern of 1s and 0s can be entered into a FF chain by forcing all the individual Q outputs to adopt a desired state simultaneously. The circuits for accomplishing this will not be shown.

Consider this chain, with all its outputs at 0.

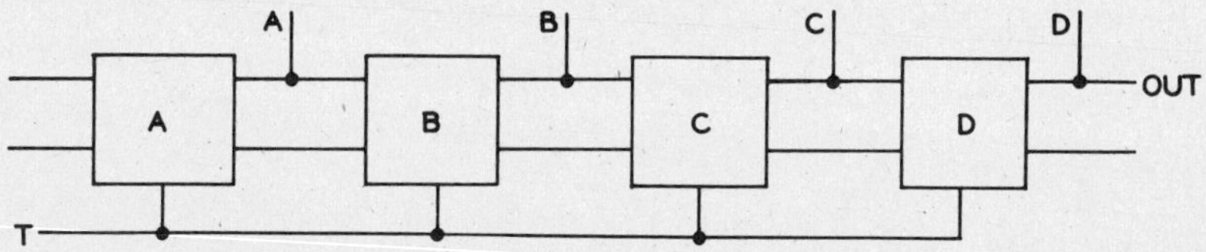


Assume it is now forced to adopt the pattern A=1 B=0 C=0 D=1. When T pulses arrive, this pattern will be passed down the line and will appear at the D output one bit at a time. In other words, the pattern comes out of the chain in serial form.

But, as it was entered all at once, it was entered in parallel form. So we have performed *parallel-to-serial* conversion. Computers require patterns of 1s and 0s in parallel form. A large binary number is a pattern of 1s and 0s, and as such all the bits must be present when the computer uses the number in calculations.

However, when transferring information from place to place, it is often better to do it in serial form. If there are 8 bits in the number, then obviously 8 wires would be needed to transfer the information to another place in parallel form. In serial form, it can be sent over one wire, one bit at a time. When using a single microwave radio link, then the information must be sent in serial form.

When data is entered into a terminal which is in a different city from a computer, then serial-to-parallel and parallel-to-serial conversion is performed continually.

In the world of human beings, we prefer to use the familiar decimal system—for the simple reason that it is familiar. In the computer world, it is easier for electronic equipment to use the binary system. Since we humans use computers and must enter information into them and they must give us answers back, then some means of going between decimal and binary is obviously needed.

**Q:** Draw four columns labelled A, B, C, and D. Write the binary numbers from 0 to 9 in these, creating ten rows. Use a fifth column labelled X and write 0 to 9 in decimal in the corresponding rows.

**A:** You should have a table like the one shown on the following page.

| A | B | C | D | X |
|---|---|---|---|---|
| 0 | 0 | 0 | 0 | 0 |
| 0 | 0 | 0 | 1 | 1 |
| 0 | 0 | 1 | 0 | 2 |
| 0 | 0 | 1 | 1 | 3 |
| 0 | 1 | 0 | 0 | 4 |
| 0 | 1 | 0 | 1 | 5 |
| 0 | 1 | 1 | 0 | 6 |
| 0 | 1 | 1 | 1 | 7 |
| 1 | 0 | 0 | 0 | 8 |
| 1 | 0 | 0 | 1 | 9 |

By examining each row, we can decide how to convert from one system to the other. First, we will convert from Binary to Decimal. 0 in decimal corresponds to all four binary inputs at 0. So if all four inputs are inverted and applied to an AND gate, the output will only go high, or to a 1, when all binary inputs are 0.

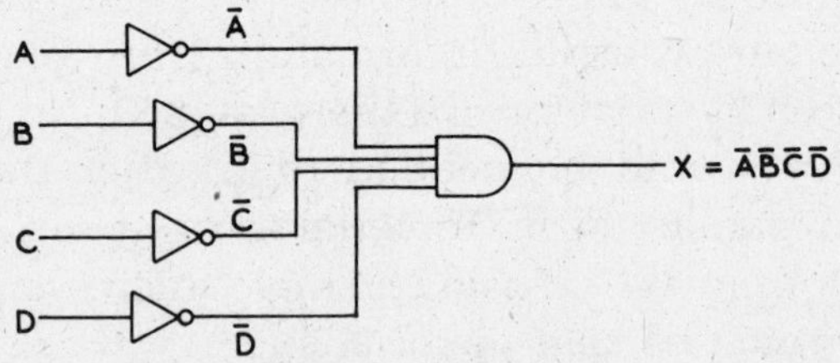


Algebraically, decimal 0 is given by $\bar{A}\ \bar{B}\ \bar{C}\ \bar{D}$. Similarly, decimal 1 is given by $\bar{A}\ \bar{B}\ \bar{C}$ D. This gives the conversion circuit.

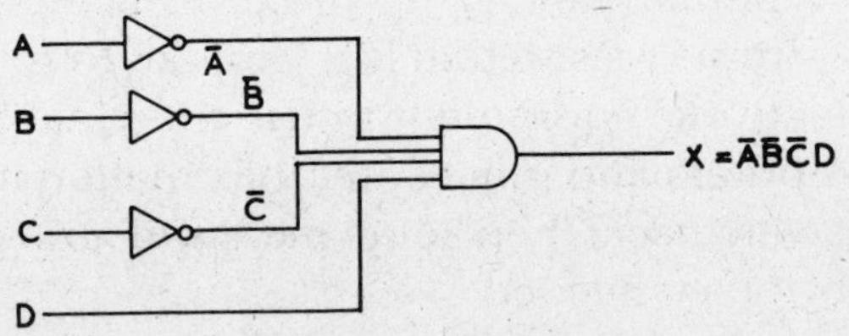


Decimal 6, for example, is $\bar{A}$ B C $\bar{D}$. This is the circuit.

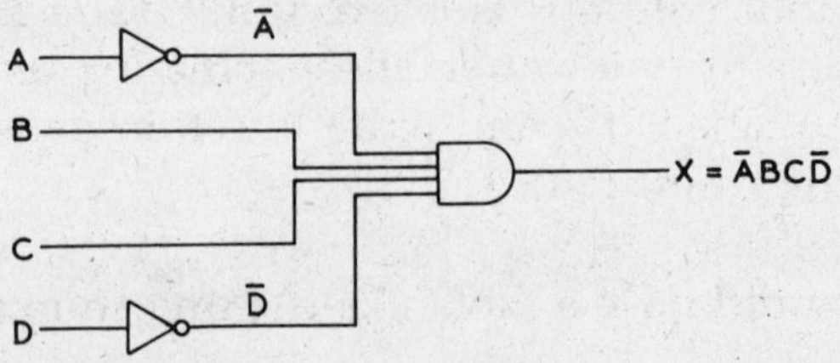

**Q:** Work out the rest of the conversions. Use algebra first, and then draw the circuit. Put all the individual circuits together to make a single Binary–Decimal converter. Hint: Only 4 inverters need be used; however, 10 AND gates are needed.

**A:** This is a Binary–Decimal converter.

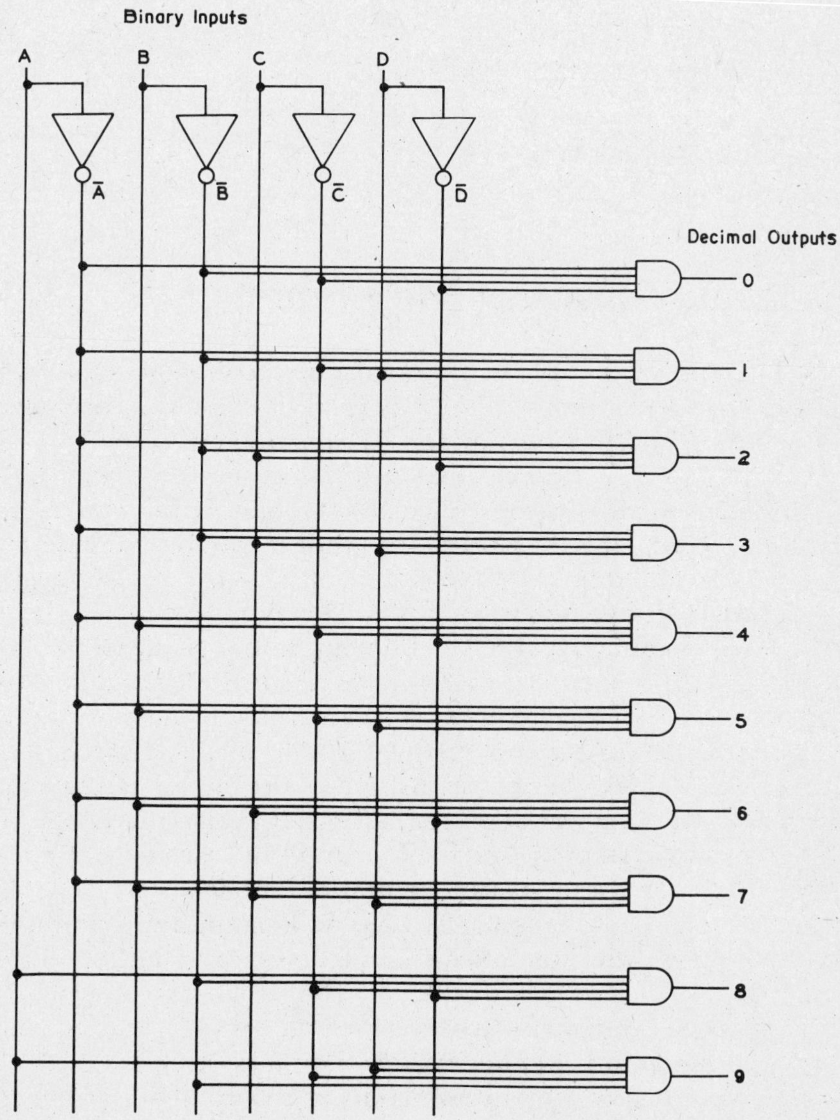

Conversion in the other direction can be effected in a similar manner. However, this circuit here shows a very simple method. The decimal inputs are normally 0, and go to a 1 successively. As they do so, they forward-bias diodes to which they are directly connected and raise the binary output line to a 1. This back-biases the other diodes on that line, and thus prevents the other lines from rising to a 1.

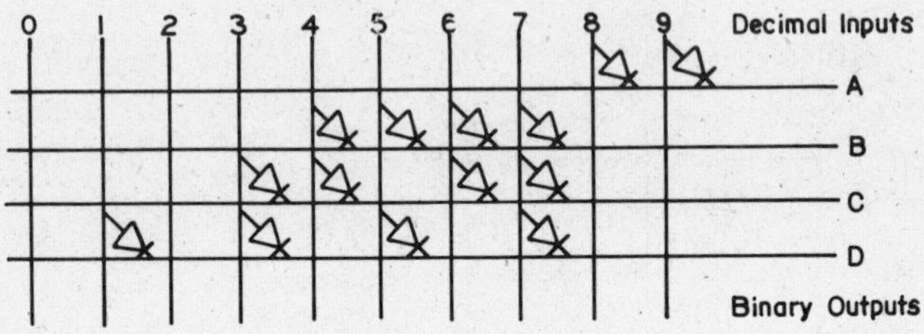


# Semiconductor Memories

When computers were in their infancy, core memories were used. These were small, ferrite, donut-shaped devices with three wires running through each one. They served as memories by being magnetized and then demagnetized by current running in one direction or another within these wires. Obviously, if eight of these donuts were required for a single "word" of eight bits, then a modest memory requirement of 8K or 8,912 bytes required an awesome volume of space to physically accommodate this requirement alone. This requirement can be realized with today's semiconductor technology with a single .8″ × .3″ integrated circuit. Core memory, therefore, is no longer used in new designs.

There are two broad classes of semiconductor memories now available. These are volatile and nonvolatile. This means that the data stored in them is either retained or lost when the power to the memory is removed. This removal of power can result from either a power failure or when the system is purposely turned off. Battery backup circuits exist to power these volatile memories under such conditions. The volatile memories are also of two types: dynamic and static RAMS. The types of nonvolatile memories are as follows: ROM, PROM, EPROM, and EEPROM.

First, let us examine the two basic fundamental operations memories undergo. A memory is either read *from* or it is written *into*. The reading operation is merely the examination of the contents of the memory, without disturbing or altering the data. As an example, a computer program might have an instruction stating that some memory location, for instance location 4387, should be read. All memories can be read from but only RAMs can be written into; that is, have their contents altered by the task of placing data into them. A RAM is really poorly named; it would be better if it were called a *read/write memory*. A RAM can be written into; and once a portion of this memory has new contents deposited into it (the writing process), the old contents or what previously was there is written over. There are two types of RAMs or read/write memories, the static and dynamic type RAMs. The static type requires that power be constantly there. If it is removed the contents will be lost. The dynamic type RAM does not have power applied to it continually, but rather in spurts. Typically, it is powered with a 1% – 2% duty cycle rather than a 100% duty cycle or being continuously ON. This part-time supplying of power to the memory still enables it to function but the power savings are significant; if the amount of RAM memory in a system is considerable and the power budget is tight, then this is an ideal candidate.

**Q:** What is the difference between a volatile and a nonvolatile memory?

**A:** A volatile memory loses its contents when power to the IC is either removed or temporarily interrupted.

**Q:** A RAM has two operations performed on it; what are these?

**A:** You can read from or write into a RAM.

**Q:** What type memories are no longer used with new designs?

**A:** Ferrite core memories.

The ROM, or read only memory is made at the factory with a set or predetermined data or bit pattern. Once this memory type is made, it cannot be reprogrammed or changed by any means. The process by which this type memory has its bit or data pattern set is called *masking*. It is rather expensive to generate a mask; therefore, ROMs have large setup costs but are cheap to mass-produce once they are made. It is this type of memory that is used in large-volume applications such as home video games, appliances, etc.

The PROM is a programmable ROM; that is, it comes from the factory "blank" and the user enters into it his or her own data pattern. This is done by programming it or applying a rather high voltage along with a programming pulse to each memory location until the capacity of the memory has been reached. The programming process is really one of either fusing (shorting) or blowing (open-circuiting) these cells. After this is done the pattern is permanent and cannot be changed. Often a paper tape reader that reads paper with little holes punched in the tape is used. It is the manner in which these little holes appear that determines the program or the data to go into the PROM.

There is an immediate problem here: namely, if you make just one mistake, your PROM reflects this; since it is permanent or unalterable, the PROM has to be discarded and is useless. This prompted the advent of the EPROM or erasable PROM. Once programmed, the EPROM is for all purposes permanent, with one small exception: If the contents need to be changed they can be. This is done by physically placing the EPROM under an ultraviolet light source for 20 minutes; as the light shines through the translucent quartz crystal glass window on top of the IC, erasure occurs. This was a great leap forward in technology; for initial development work in which programs are constantly being changed in their formative process, this is ideal. Also, low-volume production units can use these to great advantage.

The process of physically removing these memories from a piece of equipment for erasure can prove to be difficult, so the EEPROM or electrically erasable, programmable ROM was developed. This very new device uses an electrical signal to erase the contents of the memory and is proving to be very useful. The one drawback is that these memories typically required a +25V programming voltage and since most TTL systems use +5V, the problem immediately arises as to how to get additional voltage. A transformer can step up or step down AC voltage in direct proportion to the primary/secondary turns ratio, but we are dealing with DC voltages here which will not even pass through a transformer. Thus, a different approach is needed.

What is usually done is to use a DC-to-DC converter. This circuit takes a smaller voltage (+5V) and converts it to the larger voltage (+25V) by use of the principle of having a switch (transistor turning on and off rapidly) pumping current into an inductor or coil; and this is held by a capacitor, which is in essence a device best described as a "reservoir" or energy holder.

**Q:** What is the difference between an EPROM and an EEPROM?

**A:** The EPROM is erased by exposure to ultraviolet light for 20 minutes while the EEPROM is electrically erased.

**Q:** For very large-volume applications what type of memory that holds data and cannot be written into is used?

**A:** A ROM.

**Q:** What type of circuit converts one DC voltage level to another DC voltage level?

**A:** A DC-to-DC converter.

## Summary

This has been a very brief look at the simplest and most common digital circuits used to perform mathematical operations. Circuits such as these are found extensively in computers, communications, and industrial automatic equipment.

The methods of Decimal-to-Binary conversion or vice versa shown here are just two of very many. In a similar manner, many other conversions are possible between a wide variety of codes. This is important in data and information transfer. Two examples will illustrate this.

1. A data terminal keyboard uses the well-known ASCII code, which is the standard for transmitting information from place to place. But the numbers used by a computer must be in binary or BCD; hence, conversion is necessary.
2. The "supermarket" code is a pattern of lines printed on packages, and can be read optically or magnetically. This code is converted by a computer to a price which is read out on the cash register and also to a number which identifies the item and adjusts inventory records accordingly.

We only briefly covered addition and counting, and division only by 2. But a full range of arithmetic and algebraic operations are possible by extensions of these basic circuits. The common hand calculator uses these same circuits, but many times over and miniaturized to an incredible degree.

The only limit to the uses of digital circuits appears to be the imagination, and this applies also to their design. Although many circuits have been designed and will continue to be, the few included here represent a good starting point for the beginning student.

## Self-Test

1. Draw a half adder circuit.
2. What is this also called?
3. Draw a chain of 3 "T" type FFs.
4. Negative going toggle or clock pulses are applied to the above chain. Draw all the waveforms found in the chain.
5. If this is used as a divider, by how many will it divide?
6. How many FFs are needed to divide by 16, 64, 256, 4096?
7. What is a serial-to-parallel converter?
8. What is a parallel-to-serial converter?
9. Draw one of each of the above.
10. Draw a pulse train to represent the binary number 9.
11. Draw this above train in parallel form. Use 4 lines. Do not include the circuit of the converter.
12. Repeat Questions 10 and 11, using the number 14.
13. What can be used to remember a binary number?
14. Can this same item be used to remember any other numbers of information?

## Answers

1. Refer to preceding text.
2. Exclusive OR.
3. Refer to preceding text.
4. Refer to preceding text.
5. 8.
6. 4, 6, 8, 11.
7. A chain of FFs with synchronous clocking, which will accept data in serial form. When the data is fully entered into the chain, it can then be read simultaneously from all the Q outputs of the FFs.

8. A chain of FFs, exactly as above. Here the data is entered into the chain by forcing all the Q outputs into the desired state, so that the whole data is stored in the FF chain. Now by applying clock pulses, the data is made to appear in serial form at the output of the last FF in the chain.
9. Refer to preceding text.
10. Binary 9 is 1001. So a pulse train would be 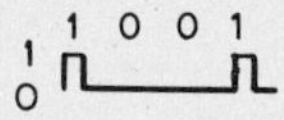

11.

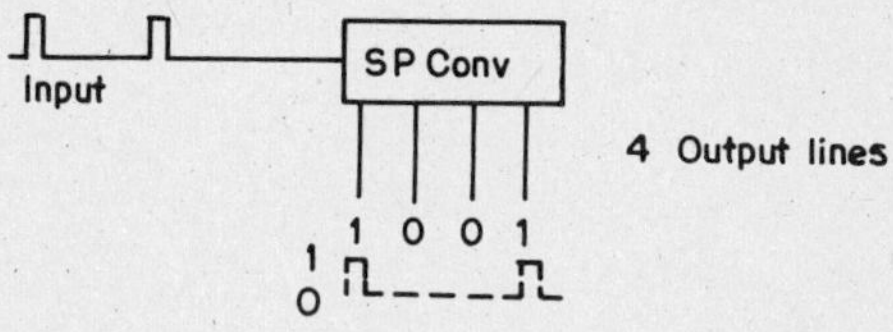


12. 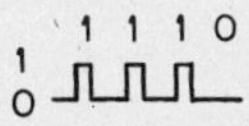


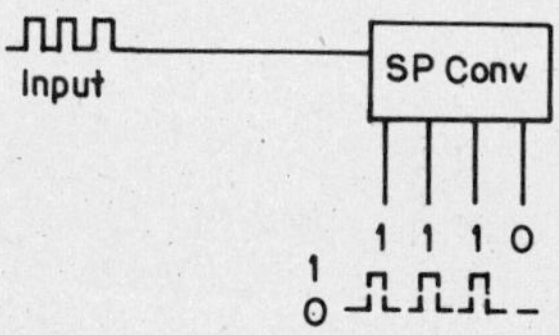


13. A chain of FFs makes a good memory, especially when used as a serial-to-parallel converter.
14. Yes, any data pattern can be remembered or stored in a chain of FFs.

CHAPTER

# 8

# Examples of Digital Circuit Applications

**OBJECTIVES:** Upon completion of Chapter 8 you will be able to do the following:

- ☐ Explain what the term "*op amp*" stands for
- ☐ Explain under what conditions an op amp's output can swing both positive and negative
- ☐ Explain how an op amp is prevented from going into saturation
- ☐ Select whether positive or negative feedback is normally avoided in amplifier use
- ☐ Explain what op amp configuration is the basis of a simple D/A amplifier
- ☐ Solve D/A converter problems
- ☐ Explain what an op amp without feedback connections is called
- ☐ Explain what type of circuit is required to make a ramp generator's ramp ideally linear
- ☐ Explain which side of a dual slope ramp voltage has a constant slope
- ☐ Explain the inherent advantage of using a dual slope integrating converter
- ☐ Explain the problem that plagues the dual slope ramp A/D converter
- ☐ Explain what the time for an A/D converter to perform its conversion function is called

- ☐ Write the purpose of a Sample/Hold module
- ☐ Explain which device helps curb the undesired effects of oscillations in digital circuitry because of slow rise times
- ☐ Use an ordinary TTL gate to make a Schmitt trigger
- ☐ Explain the advantages of using a crystal oscillator
- ☐ Explain which type of crystal is the most fragile
- ☐ Explain which type of crystal has the least piezoelectric tendencies
- ☐ Explain how a sinewave output is converted to a squarewave
- ☐ Explain why crystal oscillators used for timekeeping applications have an output that is some power of 2

THIS CHAPTER SERVES as a catch-all for a number of subjects in the digital area that were not covered in earlier chapters. This includes the following specific subjects:

1. A/D (Analog-to-Digital) and D/A (Digital-to-Analog) converters
2. PNGs (Pseudorandom Number Generators), also known as linear sequence generators
3. Signal Shaping and Conditioning
4. Crystal Controlled Oscillators
5. Transition Detectors
6. Frequency Doublers
7. Optically Coupled Devices

Because of the sophisticated nature of most of these subjects and the type of test equipment that would be required, no experiments are included in this chapter. However, you are highly encouraged to investigate these areas further and to perform experiments on your own, if you have access to such equipment.

Before A/D and D/A (Analog-to-Digital or Digital-to-Analog) converters can be discussed we must first examine a very important integral part of these converters. This is the op amp, or *OP*erational *AMP*lifier.

**Q:** What does the term op amp stand for?

**A:** Operational Amplifier.

An op amp is an amplifier with an extremely high voltage amplification capability, typically 100,000 to 1,000,000. That means that a voltage of only one-millionth of a volt could be amplified to somewhere between 0.1 and 1.0 volts. The op amp usually is operated from two supply voltages, one a positive source and the other one a negative source. Likewise, when operated like this, its output can swing either positively or negatively.

**Q:** Under what conditions can an op amp's output swing both positively and negatively?

**A:** When the op amp has both a positive and negative supply voltage.

With this high a gain it takes only a very small input signal to drive the device into saturation, or to either its maximum or minimum limit. Therefore, to prevent this, the concept of feedback is used. There are two types of feedback, positive and negative.

**Q:** How is the tendency of an op amp to saturate overcome?

**A:** Feedback is used.

Positive feedback means that as the output signal increases, this signal is fed back in a manner to allow the output to increase even further. This situation obviously results in a "run away" uncontrollable condition in an amplifier, and is to be avoided. Negative feedback, however, has the curing property of when the output increases, a portion of this signal is fed back to the input in such a manner as to cancel some of, if not all, of this excessive gain. A commercial radio employs this same principle of negative feedback for when the signal suddenly gets stronger but the volume does not get appreciably or proportionately greater. This is called AGC, or automatic gain control.

**Q:** Is it positive or negative feedback that is generally avoided in amplifier use?

**A:** Positive feedback.

Unlike some amplifiers, the op amp has two inputs. These are the noninverting input denoted by a (+) and the inverting input denoted by a (−) sign. The output and input are 180° out of phase for signals applied to the inverting input (−) and in phase for signals applied to the noninverting (+) input.

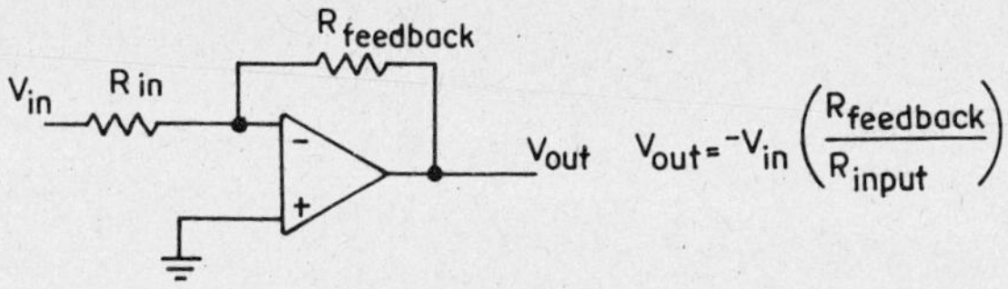


Inverting Amplifier

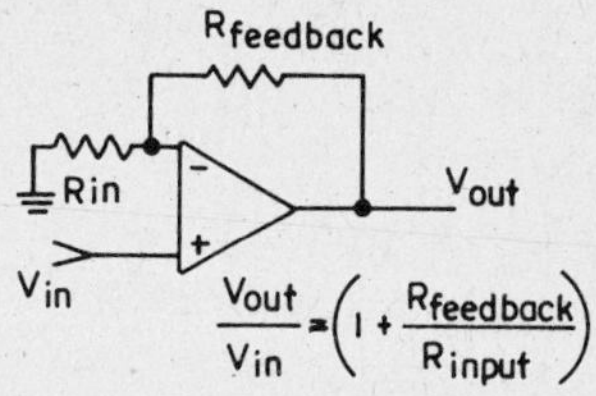


Noninverting Amplifier

Most applications use just one of the inputs and the other is grounded, either directly or through a resistor whose value is the parallel combination of the other input's resistor and the feedback resistor.

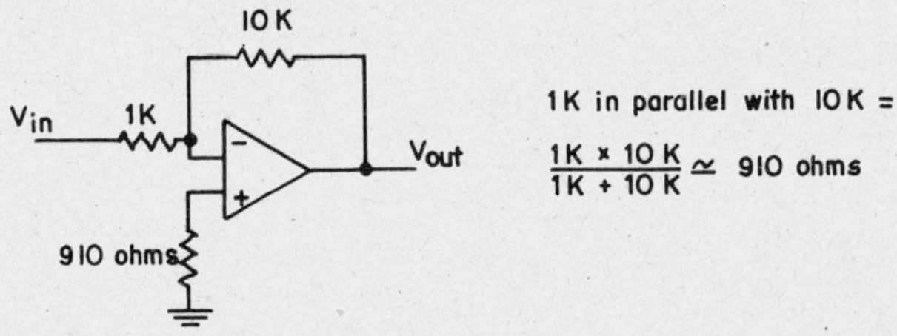


Inverting Amplifier with a gain of 10

An op amp can be used as a summing amplifier to algebraically add signals and then have their sum as the input to an op amp. This is the basis of a simple D/A converter.

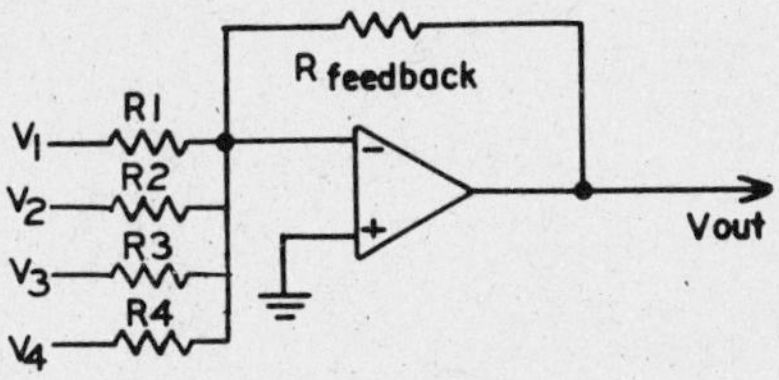


A Summing Amplifier

**Q:** What op amp configuration is the basis of a simple D/A amplifier?

**A:** An op amp used as a summing amplifier.

In actual life most D/A converters are 8 bits and above, up to as many as 18 bits. For the purposes of explanation we'll examine a 4-bit converter, realizing that these same principles can be applied to a more complex converter with more bits and better resolution. The concept of resolution means how small a number of steps on the input can be realized by the device. As an example, our 4-bit converter, like any D/A converter, will have a full-scale output of $(2^N-1)/2^N$ times the reference voltage with N being the number of bits. Since our converter is a 4-bit converter, $(2^4-1)/2^4$ or 15/16 times the reference voltage is what we will realize at full scale or when all digital ones appear at the input.

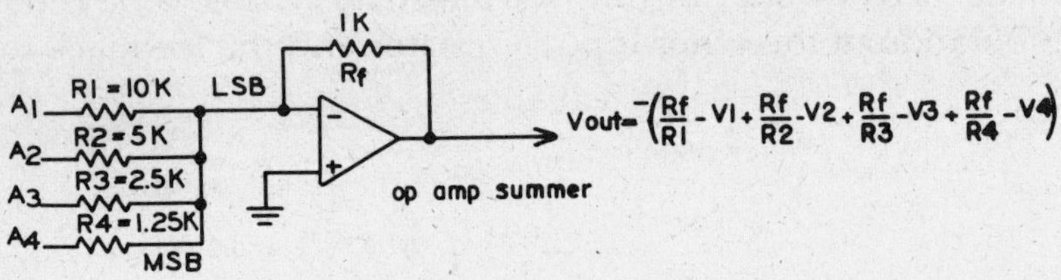


A 4 Bit D/A Converter

**Q:** If the reference voltage of an 8-bit D/A is 10 volts, what is the full-scale output voltage?

**A:** $(2^8-1)/2^8 \times 10$ volts $= 255/256 \times 10$ volts $= 9.961$ volts

Note carefully in the previous illustrations how each resistor doubles in value from the one adjacent to it. It probably is obvious to you now that as the number of bits increases in a D/A converter, the resolution likewise increases; however, for this to be meaningful you may need 0.1% or even more accurate tolerance resistors. Most D/A converters today come in one package with the resistors already built in, and they are often laser-trimmed or adjusted in that manner to realize this type of extremely accurate tolerance.

Sometimes the reference voltage is made slightly greater than, for example, 10 volts. Let us assume that the reference voltage in our previous example was 10.0039 volts instead of 10 volts. Now, if you go through the calculation for full-scale voltage you will get an exact 10V.; at a value of one half, the largest digital value you will obtain is exactly 5.000V. This is often done to insure a full-scale voltage of exactly some integer or whole number like 10 or 5 volts, etc.

**Q:** Why is the reference voltage sometimes changed from a whole number like 10.000V to 10.0039 volts?

**A:** So that when the input is at full scale or with all digital ones, the output will be an integer voltage like exactly 10.000 volts.

A/D converters are the opposite of D/A converters; they take an analog voltage as an input and change it to an output of a digital code. There are probably nine or ten types of A/D converters. This discussion will be limited to three of the most common types: 1) a ramp type; 2) a dual slope integrating; and 3) an A/D converter using D/A feedback. A 3-1/2 digit digital voltmeter (DVM) is discussed along with A/D converters in detail in the March 1982 *CQ* magazine (p. 44). The article is co-authored by Vaughn Martin and Jim McCullin.

Two circuits integral to an A/D converter are a comparator and a ramp generator. The op amp uses no feedback, so a small voltage or any logic level on its inputs will make it saturate, or go to its positive or negative extremes. These are in reality about 1 volt less than the positive and negative supply voltage.

**Q:** What is an op amp without feedback connections called?

**A:** A comparator.

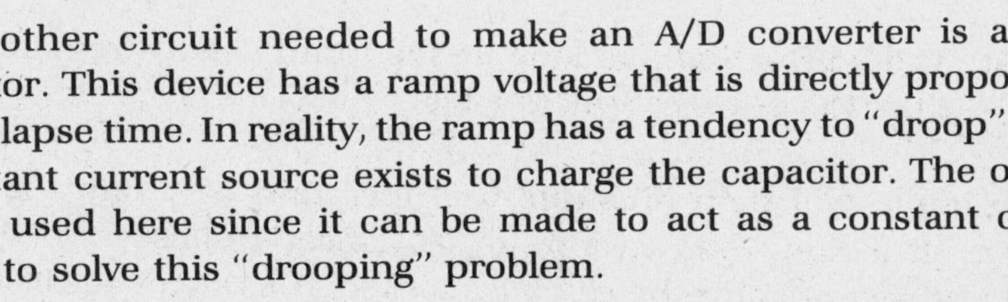


The other circuit needed to make an A/D converter is a ramp generator. This device has a ramp voltage that is directly proportional to the elapse time. In reality, the ramp has a tendency to "droop" unless a constant current source exists to charge the capacitor. The op amp can be used here since it can be made to act as a constant current source to solve this "drooping" problem.

**Q:** What type of circuit is required to make a ramp generator's ramp ideally linear?

**A:** A constant current source.

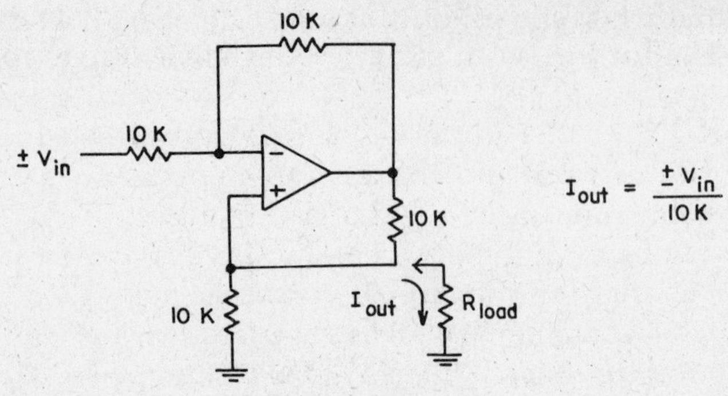


Op Amp Constant Current Source

Every practical current source has some internal resistance associated with it.

The simplest type A/D converter is the ramp type A/D as shown. The ramp type A/D has a CLEAR command that sets the binary counter to a state of all zeros. At the same time the ramp generator is applied to the comparator. The other input of the op amp comparator receives the analog input to be measured. The analog input voltage is greater than the instantaneous ramp voltage so the AND gate's output is high while being clocked. When the ramp voltage just exceeds the analog input voltage, the comparator's output goes low, making the AND gate pair go low. This disables the counter. The number of clock pulses then approximately equals the analog input voltage.

**Q:** The AND gate's output remains HIGH as long as what condition is met?

**A:** The AND is high as long as the ramp voltage remains lower than the analog input voltage.

An example will help illustrate the operation of a ramp type A/D converter. If a ramp has a slope of 0.5V/msec (millisecond or thousandth of a second) and the clock is clocked at a rate of 400,000 pulses per second (400 kHz squarewave), and the full scale of the A/D or the limit of the input is 5V, how large of a binary counter is required? Let us analyze this problem in greater detail. First, if 5 volts is full scale and it takes 0.5V/msec then the ramp generator will take 10 msec to reach 5V.

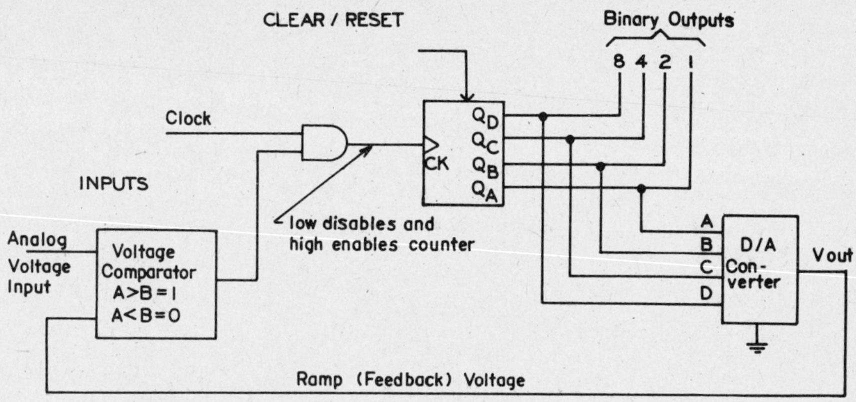


4-Bit Counter Ramp Type A/D Converter

During this 10 msec, the clock is running at a 400 kHz rate. Therefore, 10 msec is 1/100 of a second, so 1/100 of the total 400,000 squarewave pulses will be counted as a maximum (full-scale input). This means 4,000 pulses must be counted. A 12-bit counter will accommodate 4,096 pulses, so a 12-bit counter will suffice.

**Q:** If an A/D (ramp type) has a 100,000 Hz (100 kHz) clock, a 1 V/msec ramp slope, and a 10V full-scale input limit, what capability counter is required?

**A:** A 10 msec period will transpire on a full 10V input; multiply 100 kHz by 10 msec and obtain 1,000. A 10-bit counter has a 1,024-bit capacity and will accommodate this particular requirement.

A more sophisticated A/D converter is the dual-slope A/D. This type A/D converter switches between the analog input voltage and a negative reference voltage. The switch, which is a semiconductor device like a transistor and not a mechanical switch, is controlled by the MSB (most significant bit) of the counter. When the MSB is a logic 0, the analog input voltage to be measured is connected to the ramp generator as shown. When the MSB is a logic 1, the negative reference voltage is connected to the ramp generator.

The operation of the A/D converter then occurs in two distinct phases. Starting with the counter set to zero, the analog input voltage that is to be measured is switched to the input of the ramp generator. The ramp generator is then merely a linear (straight-line) ramp with a slope proportional to the input voltage $V_{in}$, as shown by line A. The comparator compares the output of the ramp with ground (0V). As stated previously, ground is a perfectly legitimate voltage level. Anyhow, the output of the comparator is a logic high. But when the A/D converter (for instance, an 8-bit converter) reaches 01111111, its next count makes it "roll over" to 10000000. When the state of the leading digit (the MSB) changes from a 0 to a 1, the semiconductor switch goes to position 2 which causes the ramp generator to be fed by the negative reference voltage ( $-V_{REF}$). Note the peak of the dual slope graph indicating this point in the transition process.

**Q:** The rising edge in the graph depicting the slope of $V_{in}/RC$ corresponds to what state of the counter's MSB?

**A:** The state of a logic 0. The falling edge corresponds to an MSB; that is, a logic 1.

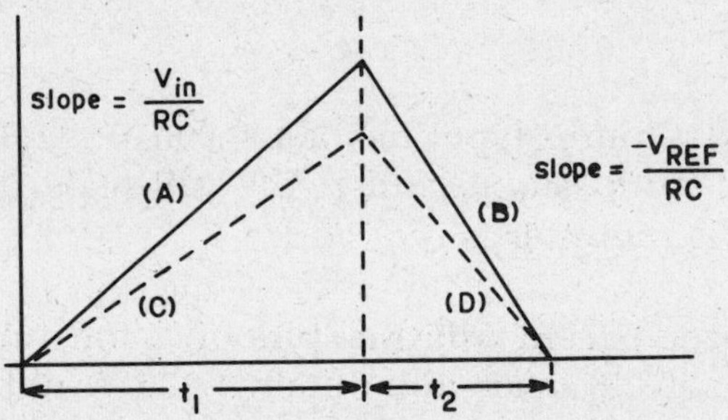


We have looked at two slopes here. The rising slope represents a variable in that the larger the analog input voltage to the A/D converter, the steeper the slope. The falling half of the dual ramp has a *constant slope* that is determined by $-V_{REF}/RC$ which is a constant nonvarying quantity. It is then the number of clock pulses recorded during this time interval ($T_2$) that represents the magnitude of the input voltage.

**Q:** Which side of the dual slope ramp has a constant slope?

**A:** The falling slope.

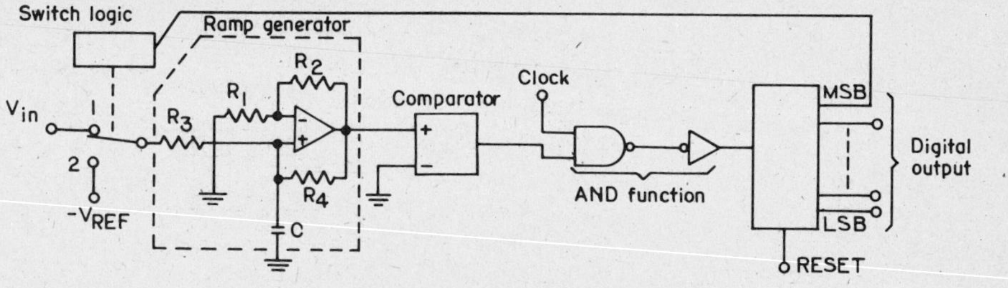


Dual slope A-D converter

This type A/D converter has the inherent advantage that it does not suffer in accuracy as the long-term accuracy of the crystal in the clock oscillator drifts or in the tolerances of the resistors and capacitors. This is quite in contrast to D/A converters with their laser-trimmed resistors, you will recall.

**Q:** What is the inherent advantage of using a dual slope integrating A/D converter?

**A:** Its accuracy is not based upon the tolerance of components or the stability of the clock.

One problem though that the dual slope ramp type A/D converter has is with the stability and linearity of the ramp generator. A solution to this is to eliminate the ramp generator and in its place use the output of a D/A converter. This feedback technique then compares the D/A output with the analog input.

**Q:** What problem plagues the dual slope ramp A/D converter?

**A:** The stability and linearity of the ramp.

**Q:** How is this problem overcome?

**A:** A D/A converter's output is fed back to the comparator and compared against the input voltage.

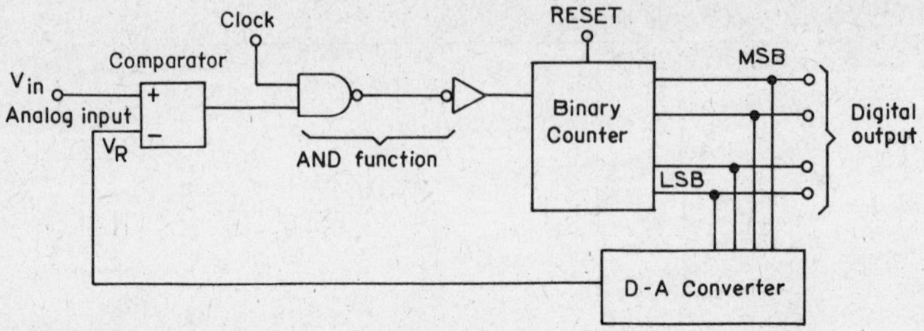


As with the dual slope ramp A/D, the D/A feedback A/D converter has its binary counter reset to a state of all zeros by the CLEAR pulse. The D/A's output is also zero, so the comparator's output is high. This allows the clock pulses to be transmitted to the counter. As the number of clock pulses increases in direct proportion with time, a stairstep (a ramp of sorts) is generated. When the analog input voltage is exceeded by the stairstep voltage, the comparator output goes low and the counter is disabled. This stops the counter and this count is a digital word representative of the analog input voltage.

**Q:** What causes the counter to be disabled?

**A:** This occurs when the stairstep voltage from the D/A's output exceeds this type of A/D converter's reference voltage ($V_R$).

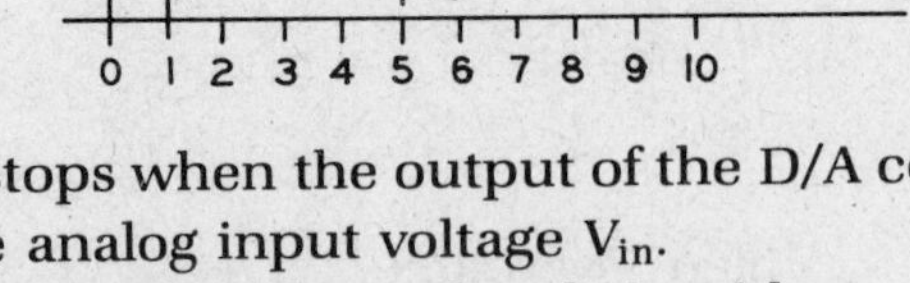


The counter stops when the output of the D/A converter $V_R$ becomes greater than the analog input voltage $V_{in}$.

One problem we will address shortly is with an analog input or even a digital input signal changing states which can cause false triggering or a gate to change logic states in an inconsistent manner. Likewise, a similar problem is apparent in A/D converters.

The time an A/D converter takes to perform its functions is called its *conversion time*. In our discussions thus far we have assumed that the analog inputs remain still or nonvarying. This is not always the case in real life.

**Q:** What is the time for an A/D converter to perform its conversion function called?

**A:** The conversion time.

One solution to this problem is to use a device called a *Sample/Hold module*. This device follows or tracks the analog input signal and then it holds an instantaneous input value constant for a specified amount of time. But let's look at what is meant by sampling. Sampling is the process by which the continuously varying analog signal is held long enough for this sample to be completed, converted, or processed by the A/D converter. A sample/hold module can be as simple as a capacitor and a switch to provide a smoothing function on erratic data.

**Q:** What is the purpose of a Sample/Hold module?

**A:** It allows continuously varying analog input signals to an A/D converter to be processed accurately despite the fact that the pattern in which the signal varies may be quite erratic.

Pseudorandom number generators (PNG), also called linear sequence generators, are formidable-sounding devices but are really not that difficult to understand. Recalling that a shift register is merely composed of cascaded or seriesed flip-flops back-to-back, it is natural to expect that their purpose is to shift a bit or a stream of bits from the input (first flip-flop's D or data input) out to the Q or $\overline{Q}$ of the last flip-flop in the chain or shift register.

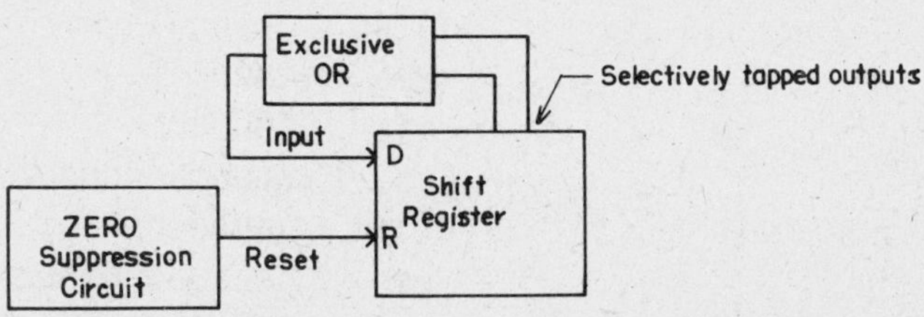


A PNG Block Diagram

What makes a PNG different is that it selectively taps off several outputs and Exclusively OR's these bits back around to the input.

Also, the output count of a PNG is a strange, randomly appearing sequence in which all numbers from a digital 1 up to $2^N-1$ appear with N being the number of bits or flip-flops composing the shift register. The order in which these numbers appear is most unusual with a recycling after the whole sequence. The pattern is repeated; but during its countdown, the sequence appears to be random but it really isn't because it repeats itself. This is why it is called a pseudo or falsely random sequence or number generator. Many modern hand-held calculators have this feature; also, video games and games of chance as well as certain sound effect generators make use of this principle.

| N | $2^N$ | $2^N-1$ | No. Bits | Feedback Connection |
|---|---|---|---|---|
| 1 | 2 | 1 | 2 | 11 |
| 2 | 4 | 3 | 3 | 011 |
| 3 | 8 | 7 | 4 | 0011 |
| 4 | 16 | 15 | 4 | 1001 |
| 5 | 32 | 31 | 5 | 00101 |
| 6 | 64 | 63 | 5 | 01001 |
| 7 | 128 | 127 | 6 | 000011 |
| 8 | 256 | 255 | 6 | 100001 |
| 9 | 512 | 511 | 7 | 0000011 |
| 10 | 1,024 | 1,023 | 7 | 1000001 |
| 11 | 2,048 | 2,047 | 8 | 00011101 |
| 12 | 4,096 | 4,095 | 8 | 01110001 |
| 13 | 8,192 | 8,191 | 9 | 000010001 |
| 14 | 16,384 | 16,383 | 9 | 000100001 |
| 15 | 32,768 | 32,767 | 10 | 0000001001 |
| 16 | 65,536 | 65,535 | 10 | 0010000001 |
| 17 | 131,072 | 131,071 | 11 | 00000000101 |
| 18 | 262,144 | 262,143 | 11 | 01000000001 |
| 19 | 524,288 | 524,287 | 12 | 000001010011 |
| 20 | 1,048,576 | 1,048,575 | 12 | 100101000001 |
| | | | 13 | 1000000000111 |
| | | | 14 | 01000000000111 |
| | | | 15 | 000000000000011 |
| | | | 16 | 0000000000101101 |
| | | | 17 | 00000000000001001 |
| | | | 18 | 000000000010000001 |
| | | | 19 | 0000000000000100111 |
| | | | 20 | 00000000000000001001 |

Note:
Where 1s occur within the "feedback connection" column, feed these bits back around through an Exclusive OR gate to the input of the shift register.

Look at the above table and refer to the feedback connection column. This indicates that in a 3-bit PNG the second and third flip-flop outputs must be tapped off (011); refer to the preceding figure too. Incidentally, when a PNG is first turned on, its output will come up in a random, unpredictable manner. When an Exclusive OR gate is used to feed the digits back to the input, the state of all logic zeros must be suppressed or it will "lock up," since no logic ones will be present to be clocked "down the line" in the shift register. Likewise, if an Exclusive NOR gate is used, the state of all logic ones must be suppressed. The circuit below will produce the strange counting sequence of 7, 3, 1, 4, 2, 5, 6 which then repeats itself.

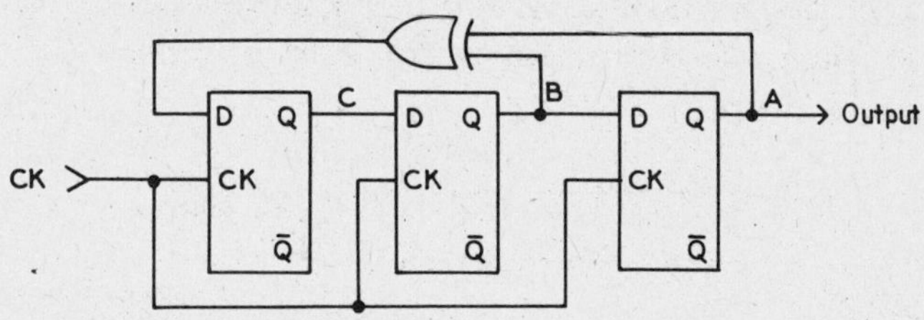


3-BIT PNG (Pseudorandom Number Generator)

**Q:** If an Exclusive-OR gate is used for feedback, what precautions need to be taken to insure the circuit will always function properly?

**A:** A scheme must be devised to suppress a possible condition in which all logic zeros occur.

**Q:** If a 4-bit PNG is used, what are the two possible manners of feedback connections? Refer to the preceding table.

**A:** The first and fourth or the third and fourth bits can be tapped.

## Signal Conditioning and Shaping

It is probably obvious to you now that digital electronics is concerned with either distinct logic levels of a 1 or 0 or with a squarewave. However, in the real world, there are numerous analog or continuously varying quantities that are input to, for example, A/D or D/A converters. What would be preferred and make life much simpler would be to have pure, clean squarewaves with sharp rise-and-fall times; or, that is, when looked at on an oscilloscope, the waveforms should appear to go straight up and down to make a true "square". This does not always happen, so undesired oscillations sometimes occur.

A method to curb this is to use a device called a *Schmitt trigger*.

**Q:** What device helps curb the undesired effects of oscillations in digital circuitry because of slow risetimes?

**A:** A Schmitt trigger.

A Schmitt trigger has a point at which it triggers or fires and another point at which it relaxes so that a slow ramp or risetime is changed into a sharp squarewave output.

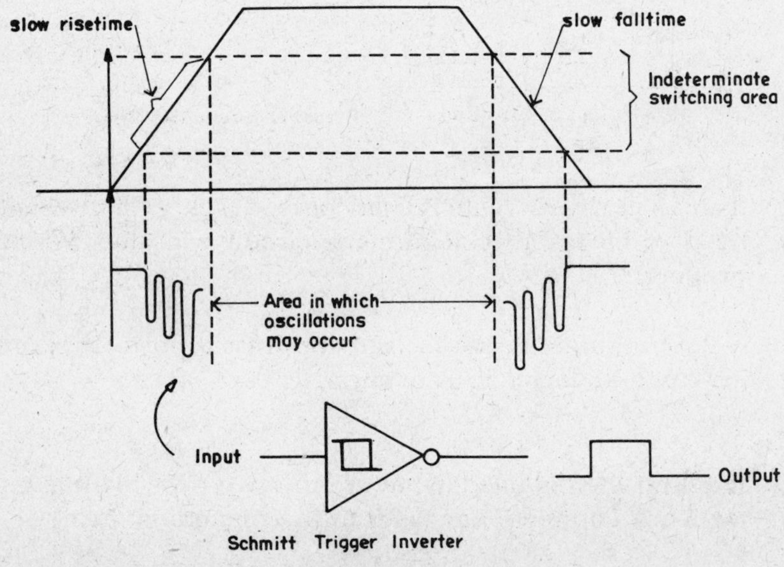


There are also ways of adding positive feedback to an ordinary gate, a TTL inverter pair is a prime example, and this makes the circuit then impervious to slow risetimes and ramps.

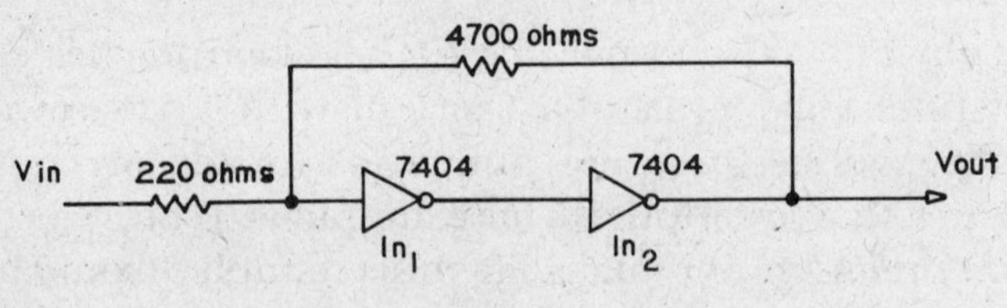

**Q:** How can a Schmitt trigger be made from an ordinary TTL gate?

**A:** Use positive feedback.

You will see this circuit used as an output for many crystal oscillators because even the overdriven sinewave still has a "curved" portion of the signal, so this method makes it really a sharp, crisp, clean squarewave. A 7413 is a typical Schmitt trigger. This is a dual 4-input NAND gate. Schmitt triggers also come in CMOS.

**Q:** Why do many crystal oscillator clocks use Schmitt trigger outputs?

**A:** This device wave shapes and sharpens up the output making it more "square" in appearance by improving rise-and-fall times.

## Crystal Controlled Oscillators

Digital wristwatches use crystals as the basis of their design. This is because oscillators built with crystals have an extremely accurate and stable output frequency.

**Q:** What are the advantages of using a crystal oscillator?

**A:** Accuracy and stability.

There are three main types of crystals. These are quartz, Rochelle salts, and tourmaline. Each of the crystals exhibits what is known as a piezoelectric effect. This means that as you apply an AC voltage across these crystals, they vibrate, or oscillate, at the AC frequency. Rochelle salts have the greatest piezoelectric effect but are the weakest and most easily breakable of the three crystal types. Tourmaline has the least piezoelectric tendencies but is very rugged. Quartz is a compromise of both, so it is used in almost all crystal oscillator applications.

**Q:** Which type of crystal is the most fragile?

**A:** Rochelle salts.

**Q:** Which type of crystal has the least piezoelectric tendencies?

**A:** Tourmaline.

A quartz crystal has one fundamental frequency at which it has maximum amplitude output. This is called its *resonant frequency*. This output is in the form of a sinewave. However, this sinewave is usually amplified to a point where it is purposely overdriven. This causes clipping and this is how the sinewave is wave-shaped into a squarewave.

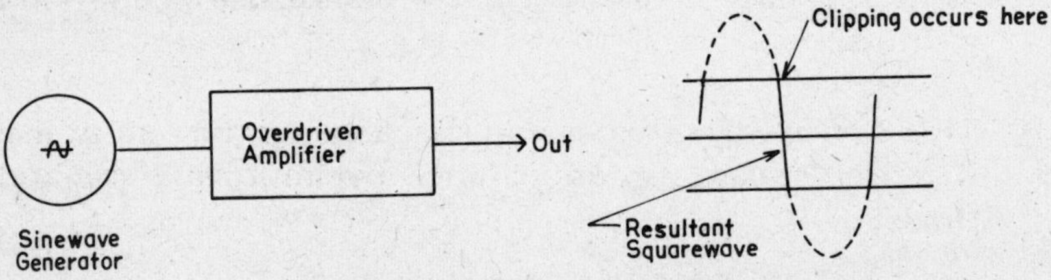


**Q:** How is a sinewave output converted into a squarewave?

**A:** It is purposely overdriven by an amplifier.

The following figure is an example of how a digital wristwatch works:

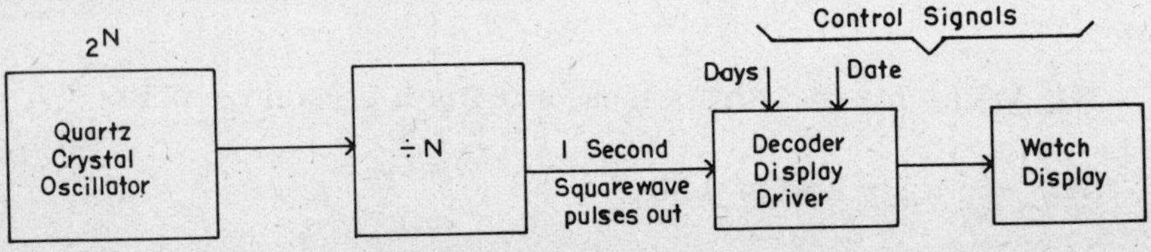


Note the frequency of the crystal. It is 32.768 kHz or $2^{15}$ so that after being divided down by 15 flip-flops or actually two counters in this case, the output is exactly one second.

**Q:** Why do crystal oscillators destined for timekeeping applications have an output that is some power of 2?

**A:** They can then be divided down exactly to one second.

## Transition Detectors

Now that the importance of the sharp, clean, crisp squarewave has been established from the discussions on Schmitt triggers, another important aspect of squarewaves that is probably already obvious to you is the detection of when a digital signal changes from a logic 1 to a logic 0. The following circuit detects this transition by having the input inverted an odd number of times, 3, 5, 7 . . . etc. This output and the input (or the original signal) are ANDed together. What happens is that B is inverted and delayed by the number of gate propagation delays. The cumulative total of these propagation delays is the width of the detected signal; refer to the following figure.

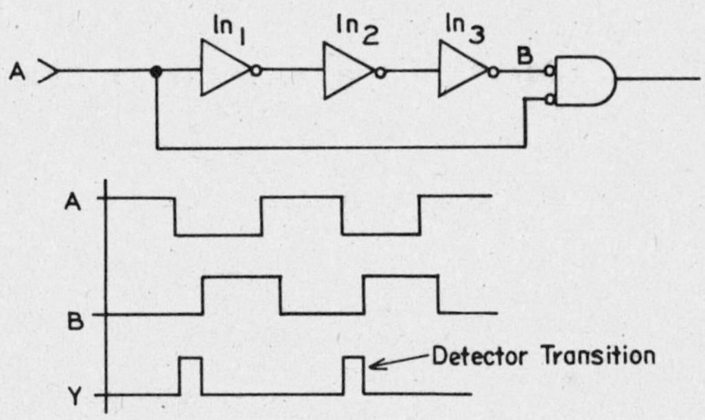


**Q:** If there is a change from a logic 1 to a logic 0, by what manner is the signal delayed?

**A:** It is delayed by the propagation delay of an odd number of gates.

If the width of the pulse indicating this 1 to a 0 transition is insufficiently wide, a capacitor may be placed from output to input of the middle inverter.

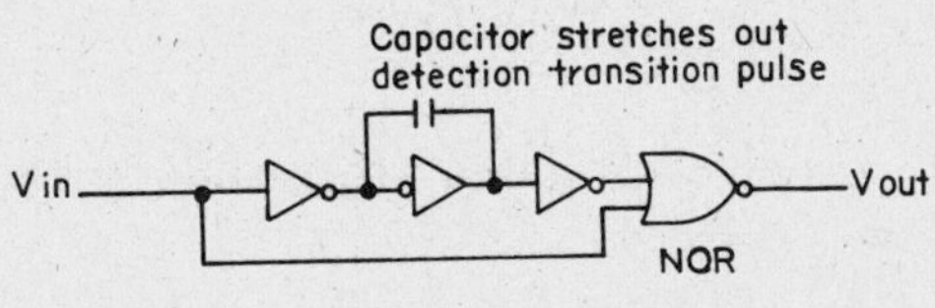

**Q:** How can the transition pulse width be widened?

**A:** A capacitor may be placed between the input and output of the middle inverter in the preceding circuit.

If a transistion detector is desired to detect the transistion from a logic 0 to a logic 1, then the NOR gate in the preceding circuit may be replaced by a NAND gate.

**Q:** What effect does swapping the AND gate with the NAND gate have in the previous circuit?

**A:** It enables the detection of a transition from a logic 0 to a logic 1.

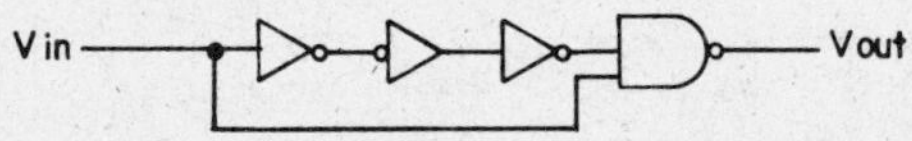


## Frequency Doublers

The circuit in the following illustration is a frequency doubler. The reader is encouraged to draw a timing diagram to verify this.

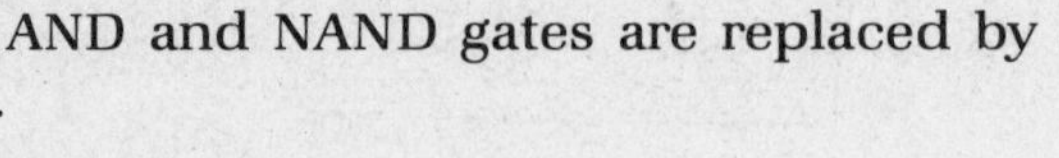


**Q:** How does the frequency doubler differ from the two previous transition detector circuits?

**A:** The AND and NAND gates are replaced by an Exclusive OR gate.

## Optically Coupled Devices

These innovative devices combine the properties of optics with the world of digital electronics. They are used almost exclusively for signal isolation purposes. But before discussing optically coupled devices, let's investigate what signal isolation is and why it is needed in digital electronics.

**Q:** What is the primary application of optically coupled devices?

**A:** In signal isolation applications.

## Signal Isolation

When attempting to interface an AC signal with a DC control signal, isolation will be required. As an example, if you wish to use a low current switch to control a motor that requires a whopping 100 amperes, such as is found in an automobile starter, the problem arises from the inability of the ignition switch to handle more than approximately 10 amperes. A relay can be used. This device is an electromechanical switch with very large, heavy copper wire windings that can carry the 100-ampere motor requirements. It is energized or made to close, though, by a much smaller DC signal; therefore, isolation from the starter motor to the starter relay is accomplished by use of a relay.

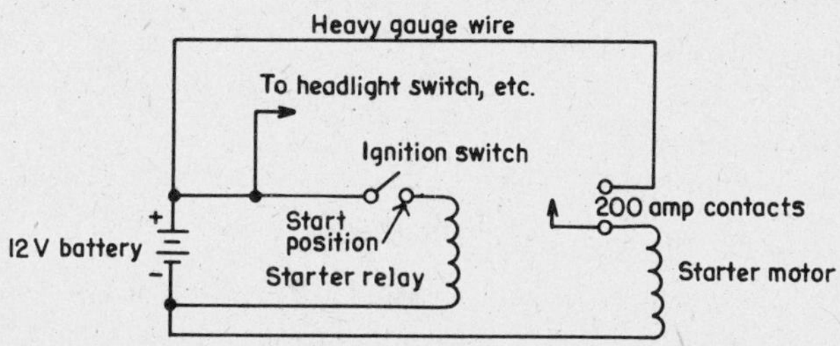


**Q:** What device acts as an electromechanical switch and allows large amounts of current to pass through its coils, when closed, and is controlled by a much smaller signal?

**A:** A relay.

Two other reasons for using some means of signal isolation are: 1) if there are electrical noise problems due to the interconnection of two systems; or 2) if the earth ground should not be considered a truly good ground because IR drops from currents in this system present a problem. Electrical noise is a phenomenon of conductive and radiated (sent out through the air) electromagnetic interference. These are the spikes on an AC line that are generated when an electrical appliance is used, and the effects are the "snow" on a TV set that suddenly goes away when the appliance is turned off.

**Q:** What is one of the two reasons to use signal isolation?

**A:** To avoid using a poor signal ground system or to help eliminate electrical noise.

Relays, as mentioned, are one method of accomplishing signal isolation, but they typically require more current to actuate them than most digital ICs can supply.

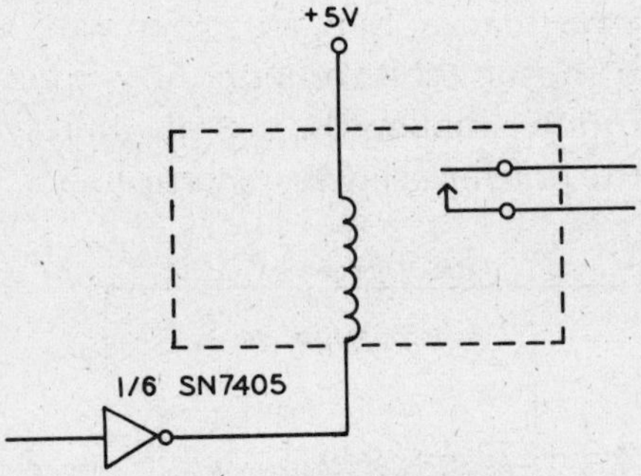


Open-collector Inverter

The answer to this problem is to use buffered or open collector TTL devices. Another method, other than relays, is to use optoelectronic isolators.

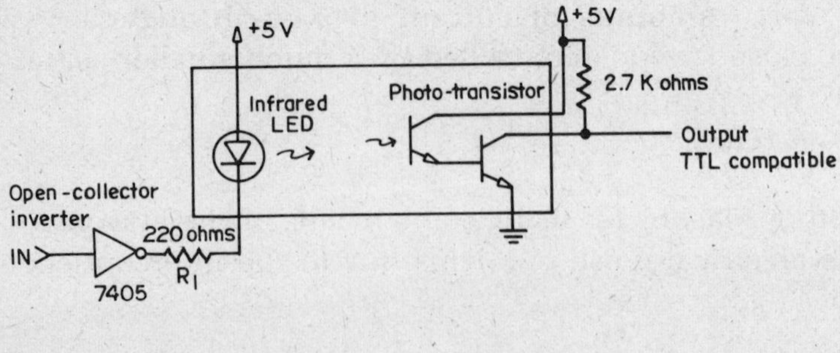

**Q:** How can the problem of insufficient output current capability in most digital ICs be overcome?

**A:** Use open collector or buffered TTL gates.

Optically coupled devices usually consist of some light-emitting source and a photodetector or a light receiver. Typically, an open collector inverter drives the LED. The photodetector is very sensitive to these light rays which are in the infared region.

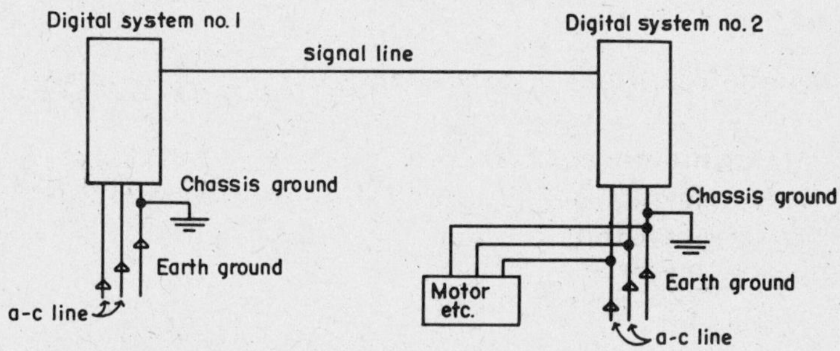


If the two earth grounds are from different sources, or if there is noise-producing equipment using the same earth ground, noise problems may exist when interconnecting signals between the two systems.

**Q:** What are the two main parts of an optically coupled device?

**A:** A photodetector and a light emitter.

Some light emitters and photodetectors are in the form of separate devices outside of an IC package. For it to function properly, they must be impervious to ambient visible light around them. This is why they emit and respond to light in the infrared region.

**Q:** Why are optically coupled devices made to respond just to infrared light?

**A:** Ordinary ambient light would cause false triggering.

Optically coupled devices, though, suffer from three distinct problems: 1) as with relays that were difficult to drive, these devices are limited to the current capability of the output transistor; 2) the output voltage is not 0 volts, as in the case of relays, but some fraction of a volt representing the saturation voltage of the output transistor or transistors; and 3) there is an obvious polarity requirement on the output. For this reason AC control is not possible with this device directly. There are devices, though, that have, for example, triac outputs which can switch AC signals. A triac is a device that allows alternating current to pass in each direction and rectifies or blocks a portion of each half of both cycles. A triac is an ideal AC motor controller or a light dimmer.

**Q:** Name the three disadvantages of optically coupled devices.

**A:** 1) output current limitations; 2) the output does not go all the way to 0 volts; 3) the output has a polarity requirement unless other provisions are made.

# Glossary

**AND-GATE.** A gate in which if all inputs are high, a high output will result.

**Astable Multivibrator.** A semiconductor device that is continuously running and produces squarewaves. This book deals predominantly with the bistable multivibrator type or flip-flop.

**Asynchronous.** An operation that is unclocked or untriggered.

**BCD (Binary Coded Decimal).** A form of numeric representation which expresses each decimal digit as a 4-bit binary nibble. This makes the driving of displays much easier. For example, a number like $3916_{10}$ would be represented as follows:

| | | | | |
|---|---|---|---|---|
| decimal: | 3 | 9 | 1 | 6 |
| binary: | 0011 | 1001 | 0001 | 0110 |

**Bistable Multivibrators.** As common practice has evolved, they are now called flip-flops. A *flip-flop* is a semiconductor device whose output changes logic states when the proper input conditions are met.

**Bit.** This term is a contraction of the words *b*inary dig*it*. For example, the binary number $101101_2$ has six binary digits, or bits.

**Boolean Algebra.** This symbolic logic is an unusual type of algebra in which letters and symbols are used to represent statements and their logical connections. This algebra's inventor, George Boole, originally assigned each variable in an equation two values, true and false. Its use has evolved, and today Boolean algebra represents the convention of looking at gates and equating them to open or closed relays or switches, and the logic functions that correspond to this.

**Buffered.** This refers to a special gate with an output that can source (drive) or sink (receive) an extra amount of current, more than the usual unbuffered devices.

**Clock.** A term synonymous with squarewave triggering.

**CMOS (Complementary Metallic-Oxide Silicon).** Another logic family in which circuit functions are realized with nonbipolar or MOS (metallic-oxide silicon) transistors. N and P channel MOS transistors are combined for the complementary effect.

**Combinational Logic.** Logic based on the assumption that a certain set of input conditions will cause a predictable output. The mere presence of the inputs in either a high or low state causes the output to be at a prescribed state. The assumption is made that these inputs are not continuously changing.

**"D" Type Flip-Flop.** A single data input device in which the output corresponds to the logic state of its input when it receives a pulse.

**Digital Electronics.** The broad branch or discipline within the field of electronics dealing with the assignment and manipulation of intelligible data and codes as a function of circuit devices or design techniques in which one of two possible states are assigned. These two states correspond to a device, such as a transistor, either being *on* (conducting) or *off* (not conducting). These circuits operate like a switch and can be designed to make logical decisions. The two assumed states, referred to as logic states, are logic "1" and logic "0".

**DTL (Diode-Transistor Logic).** An antiquated logic family only rarely used in designs today. DTSs use transistors and diodes, with their inherent characteristic of blocking and permitting current flow, to perform simple logic functions such as gates.

**Exclusive OR.** This special gate can have a high output only if its inputs are not either all high or all low.

**Fan In.** The number of inputs that can be connected to a logic input circuit.

**Fanout.** The number of gates within the same logic family that can be driven by a like gate. For example, if a TTL gate can drive ten other TTL gates then it has a fanout of 10.

**Floating Input.** An input in which the voltage level is allowed to remain or drift into a range that is neither a logic 1 nor a logic 0 but somewhere in between. This is a forbidden area in which to reside, and often causes hard-to-trace problems.

**Gates.** These are the most predominant class of digital semiconductor circuits. Gates are capable of assuming or being assigned one of the two possible binary states of a logic 1 (high) or logic 0 (low). Gates are further characterized by the ability to have certain input conditions produce a predictable output state.

**Glitch.** An unwanted pulse that can cause false outputs. The possible causes are numerous, but commonly it is the simultaneous changing of inputs on a multiple input gate in which one input experiences a low-to-high logic level transition while another input at the same time undergoes a high-to-low transition. *Note*: A NAND, NOR, and Exclusive NOR gate also exist and have truth tables that are just the opposite of the AND, OR, and Exclusive OR gates respectively.

**Hysteresis.** As applied to Schmitt trigger switching action, this specifically means that there is a lag time from switching states, either from high-to-low or vice versa, during which either a predefined high or low voltage level must occur before switching action transpires. This prevents intermediate state switching or oscillations in which spurious switching occurs in a region between a truly legitimate logic high or low.

**Integrated Circuit.** Semiconductor device with a combination of interconnected circuit elements inseparably placed on a continuous substrate or single chip of silicon, with small, protruding, regularly spaced metallic pins exactly 0.1 inch apart for connection purposes. These integrated circuits, or ICs as they are more commonly known, are usually encased in plastic or ceramic material. All of the previously mentioned gates are usually realized in the form of an integrated circuit.

**Inverter.** A single input device which sets the output opposite to the input state.

**JK Flip-Flop.** This type of flip-flop has two control inputs called J and K. The logic states of J and K control the output.

**Karnaugh Map.** A graphic method for simplifying and minimizing logic circuits.

**Latch or Data Latch.** A special type of flip-flop in which a logic state is retained until a pulse is applied to one input, causing the complementary data to be stored. These devices are used for data storage as opposed to counting and logic functions. This is the most rudimentary form of a memory device.

**Logic Family.** A collective grouping of integrated circuits with the common factor of design with respect to circuit functions with active devices (transistors).

**Logic 1 or Logic High.** The presence of voltage at a predetermined level at the output or input of a semiconductor logic device.

**Logic 0 or Logic Low.** The absence of voltage or nearly ground level at either the input or output of a semiconductor logic device.

**Lower Triggering Point.** The voltage level which constitutes a legitimate logic low, above which no triggering action or logic states should change until an upper triggering point is reached.

**LSB (Least Significant Bit).** The bit, or binary digit, to the far right; it is always equal to 1 if a 1 is there, and equal to zero if a binary 0 is there.

**LSD (Least Significant Digit).** The right-most digit in a grouping of bits or words.

**Monostable Multivibrator.** A semiconductor device that puts out a single pulse in the form of a squarewave.

**MSB (Most Significant Bit).** In a binary number, this refers to the left-most digit. This digit is twice the value of the one to its right, the next to MSB; and the next MSB is twice the value of the digit to its right, etc.

**MSD (Most Significant Digit).** The left-most digit in a grouping of bits or words.

**Multivibrator.** A generic term describing a number of different square-wave generators or square pulse generators. Some multivibrators are continuously running, others only put out a single pulse, and others merely change states when they are themselves pulsed.

**Numbering Systems.** The system by which we are able to count and relate quantitative data. The numbering system we are most accustomed to is the decimal numbering system which is based on the number ten. Other numbering systems, for example, are the *hexadecimal, octal*, and *binary* system which are based on the numbers 16, 8, and 2 respectively. The largest number expressable by a single digit in any numbering system is one less than that number itself. As an example, in our decimal system we can express "9" as the single larges single digit number. This is $10-1=9$. Likewise, with the binary system $2-1=1$, so a 1 is the largest single digit expressable number with 0 also a legitimate number.

**OR-GATE.** A gate in which if any one of the inputs is high, a high output will result.

**Oscillations.** In digital electronics, unwanted switching action caused by either unpredictable switching transistions or induced changing of logic states caused by electrical noise. In any event, this is to be avoided and is done so by the use of a Schmitt trigger.

**Parallel Loading (Broadside Loading).** When the data is applied to the "D" inputs of the flip-flops. It is also often "gated in" or applied to the input of a multiple-input gate with an enable and inhibit signal on one of the inputs to the gate. This enables controlled loading, or loading under certain specific conditions.

**Propagation Delay.** The very small amount of time required for the output to change in response to a change in input. This is typically 15 to 20 nanoseconds, or 15 to 20 billionths of a second for TTL logic.

**Pull Up Resistor.** A resistor with one side tied to a positive voltage level at or above a potential larger than that which constitutes a logic high; the other end of the resistor is tied to an input on a gate. This insures that the gate's input will always be high until another logic device goes low and in turn pulls this resistor and the gate's input down to a logic 0 voltage level. The pull up resistor convention is helpful in electrically noisy environments in which intermittent glitches could cause false triggering or transition states.

**Q and $\overline{Q}$.** The two outputs on a flip-flop. These two outputs are always in opposite states if the flip-flop is operating properly.

**Register.** A group of memory elements (usually flip-flops) operating in unison to perform a specific task. This task may simply be to hold data temporarily, which is a function of a *buffer register*.

**Reset.** Another control on a flip-flop that, like the Set, controls the $\overline{Q}$ output under like conditions.

**RTL (Resistor-Transistor Logic).** An antiquated logic family in which resistors and their placement in networks, with subsequent voltage dividing effects and cascading of transistor amplification stages, perform logic functions.

**Sequential Logic.** Logic in which a certain predefined set or sets of events have first to transpire before the expected end-result will occur. A continuously running squarewave is often used as the driving force so that the experiments based on this logic transpire within a predictable amount of time. The first three chapters of this book deal solely with the combinational logic.

**Serial Loading.** When the data is in a serial form and is applied to the "D" input of the first flip-flop in the chain of flip-flops making up the shift register. The data is propagated down the line, bit by bit, on each clock input of the flip-flops. These clock signals can be applied to the first clock input of the shift register. Its output (Q) is then applied to the successive "D" input whose "Q" is applied to the successive "D" input of the next flip-flop, etc. right on down the line. This is called a *ripple counter* because the data "ripples" or propagates down the line serially. Shift registers also exist with only one clock signal applied to all of the clock inputs of the flip-flops simultaneously. This is called *synchronous counting*. Your study of digital electronics will involve microprocessors, and within LSI chips is a device called a *program counter*. This special register combines the features of both a buffer register and a counter. Its purpose is to count and be incremented by one count on each clock cycle, allowing the program to be executed.

**Set.** A control on a flip-flop that when active or in the asynchronous mode, controls one of the flip-flop's two outputs, usually the Q output.

**Shift Register.** The most common type of register. It accepts data and shifts it to the right or left, or adds 1 to the data. Another type of register that merely has a logic 1 circulated around through all its outputs is called a *ring counter*.

**Synchronous.** The clocking of circuits in which the responses and events in general that occur are based upon certain squarewaves or numbers of squarewaves present.

**Timing Diagram.** A diagram drawn on grid or graph paper to exact scale, in which the relationship of occurring events are compared to the stimulus or squarewave generator driving the circuits.

**Triggering.** The process of supplying a pulse to a logic device. Most logic devices we will study from here on will be triggered by a pulse or by a continuous stream of squarewaves. There are two types of devices in logic that are subject to having their outputs change from an input change. These are *edge* and *level* sensitive devices. The first must see a *rising* or a *falling* squarewave edge, while level sensitive devices only respond to logic levels. The devices that trigger on the rising edge have no bubble or small circle on the triggering input, while devices that trigger on the falling edge are represented by this bubble or small circle, generally denoting inversion.

**Truth Table.** This is a systematic and thorough means by which all possible input conditions to a gate are listed along with the corresponding output that will result from each particular input condition.

**TTL (Transistor-Transistor Logic).** This is an example of a logic family with gates and more complex logic functions, realized by the back-to-back placing of bipolar transistors.

**Upper Triggering Point.** The voltage level which constitutes a legitimate logic high. Voltage levels below this should have no effect on the switching action. An input signal must reach this level and go low again before the digital or logic device changes states. The time during this transistion state is a function of the amount of hysteresis in the circuit.

**Word.** A collective grouping of bits. The length of a word is up to the circuit designer, although the most popular word lengths are 4 bits (called *nibbles*), 8 bits (called *bytes*), and 16 bits long.

# Index